Energy, Climate and the Environment

Series Editor: **David Elliott**, Emeritus Professor of Technology Policy, Open University, UK

Titles include:

Jo Abbess
RENEWABLE GAS
The Transition to Low Carbon Energy Fuels

Giorel Curran
SUSTAINABILITY AND ENERGY POLITICS
Ecological Modernisation and Corporate Social Responsibility

Declan Kuch
THE RISE AND FALL OF CARBON EMISSIONS TRADING

Claire Dupont and Sebastian Oberthür (*editors*)
DECARBONISATION IN THE EU
Internal Policies and External Strategies

Espen Moe
RENEWABLE ENERGY TRANSFORMATION OR FOSSIL FUEL BACKLASH
Vested Interests in the Political Economy

Manuela Achilles and Dana Elzey (*editors*)
ENVIRONMENTAL SUSTAINABILITY IN TRANSATLANTIC PERSPECTIVE
A Multidisciplinary Approach

Robert Ackrill and Adrian Kay (*editors*)
THE GROWTH OF BIOFUELS IN THE 21ST CENTURY

Philip Andrews-Speed
THE GOVERNANCE OF ENERGY IN CHINA
Implications for Future Sustainability

Gawdat Bahgat
ALTERNATIVE ENERGY IN THE MIDDLE EAST

Ian Bailey and Hugh Compston (*editors*)
FEELING THE HEAT
The Politics of Climate Policy in Rapidly Industrializing Countries

Mehmet Efe Biresselioglu
EUROPEAN ENERGY SECURITY
Turkey's Future Role and Impact

Jonas Dreger
THE EUROPEAN COMMISSION'S ENERGY
A Climate for Expertise?

Beth Edmondson and Stuart Levy
CLIMATE CHANGE AND ORDER
The End of Prosperity and Democracy

David Elliott (*editor*)
NUCLEAR OR NOT?
Does Nuclear Power Have a Place in a Sustainable Future?

Neil E. Harrison and John Mikler (*editors*)
CLIMATE INNOVATION
Liberal Capitalism and Climate Change

Antonio Marquina (*editor*)
GLOBAL WARMING AND CLIMATE CHANGE
Prospects and Policies in Asia and Europe

Espen Moe and Paul Midford (*editors*)
THE POLITICAL ECONOMY OF RENEWABLE ENERGY AND ENERGY SECURITY
Common Challenges and National Responses in Japan, China and Northern Europe

Marlyne Sahakian
KEEPING COOL IN SOUTHEAST ASIA
Energy Consumption and Urban Air-Conditioning

Benjamin K. Sovacool
ENERGY & ETHICS
Justice and the Global Energy Challenge

John Vogler
CLIMATE CHANGE IN WORLD POLITICS

Energy, Climate and the Environment
Series Standing Order ISBN 978–0–230–00800–7 (hb)
978–0–230–22150–5 (pb)

You can receive future titles in this series as they are published by placing a standing order. Please contact your bookseller or, in case of difficulty, write to us at the address below with your name and address, the title of the series and the ISBNs quoted above.

Customer Services Department, Macmillan Distribution Ltd, Houndmills, Basingstoke, Hampshire RG21 6XS, UK.

Climate Change in World Politics

John Vogler
Professorial Research Fellow in International Relations, Keele University, UK

© John Vogler 2016
Softcover reprint of the hardcover 1st edition 2016 978-1-137-27340-6
All rights reserved. No reproduction, copy or transmission of this publication may be made without written permission.

No portion of this publication may be reproduced, copied or transmitted save with written permission or in accordance with the provisions of the Copyright, Designs and Patents Act 1988, or under the terms of any licence permitting limited copying issued by the Copyright Licensing Agency, Saffron House, 6–10 Kirby Street, London EC1N 8TS.

Any person who does any unauthorized act in relation to this publication may be liable to criminal prosecution and civil claims for damages.

The author has asserted his right to be identified as the author of this work in accordance with the Copyright, Designs and Patents Act 1988.

First published 2016 by
PALGRAVE MACMILLAN

Palgrave Macmillan in the UK is an imprint of Macmillan Publishers Limited, registered in England, company number 785998, of Houndmills, Basingstoke, Hampshire RG21 6XS.

Palgrave Macmillan in the US is a division of St Martin's Press LLC,
175 Fifth Avenue, New York, NY 10010.

Palgrave Macmillan is the global academic imprint of the above companies and has companies and representatives throughout the world.

Palgrave® and Macmillan® are registered trademarks in the United States, the United Kingdom, Europe and other countries.

ISBN 978-1-137-27343-7   ISBN 978-1-137-27341-3 (eBook)
DOI 10.1057/9781137273413

A catalogue record for this book is available from the British Library.

A catalog record for this book is available from the Library of Congress.

Typeset by MPS Limited, Chennai, India.

Contents

List of Figures and Tables vi
Series Editor's Preface vii
Acknowledgements ix
List of Abbreviations xi

1 Introduction 1
2 Framing and Fragmentation 13
3 The UNFCCC Regime 35
4 Interests and Alignments 60
5 The Pursuit of Justice 86
6 Recognition and Prestige 108
7 Structural Change and Climate Politics 131
8 Conclusion 157

Notes 178
References 186
Index 202

List of Figures and Tables

Figures

2.1	Contribution of sectors to GHG emissions 2010	20
3.1	World energy-related CO_2 emissions	42
3.2	Organigram, UNFCCC regime	56
5.1	The EU burden-sharing agreement 2008	95
5.2	National EU emissions 2012	96
5.3	Historic cumulative emissions 1850–2030	97
5.4	Remaining carbon space (entitlements)	100
5.5	Distribution of per capita emissions by country	101
7.1	GDP growth for major economies	145
7.2	Indicators of structural change for EU, BASICs and US 1990, 2000 and 2010	147
8.1	Global trends in renewable energy investment 2004–13	163
8.2	Emissions gap to close 2015–50	165

Tables

3.1	Chronology of the UNFCCC regime	38
7.1	The changing context of the climate regime	134

Series Editor's Preface

Concerns about the potential environmental, social and economic impacts of climate change have led to a major international debate over what could and should be done to reduce emissions of greenhouse gases. There is still a scientific debate over the likely *scale* of climate change, and the complex interactions between human activities and climate systems, but global average temperatures have risen and the cause is almost certainly the observed build-up of atmospheric greenhouse gases.

Whatever we now do, there will have to be a lot of social and economic adaptation to climate change – preparing for increased flooding and other climate-related problems. However, the more fundamental response is to try to reduce or avoid the human activities that are causing climate change. That means, primarily, trying to reduce or eliminate emission of greenhouse gasses from the combustion of fossil fuels. Given that around 80 per cent of the energy used in the world at present comes from these sources, this will be a major technological, economic and political undertaking. It will involve reducing demand for energy (via lifestyle choice changes – and policies enabling such choices to be made), producing and using whatever energy we still need more efficiently (getting more from less), and supplying the reduced amount of energy from non-fossil sources (basically switching over to renewables and/or nuclear power).

Each of these options opens up a range of social, economic and environmental issues. Industrial society and modern consumer cultures have been based on the ever-expanding use of fossil fuels, so the changes required will inevitably be challenging. Perhaps equally inevitable are disagreements and conflicts over the merits and demerits of the various options and in relation to strategies and policies for pursuing them. These conflicts and associated debates sometimes concern technical issues, but there are usually also underlying political and ideological commitments and agendas which shape, or at least colour, the ostensibly technical debates. In particular, at times, technical assertions can be used to buttress specific policy frameworks in ways which subsequently prove to be flawed.

The aim of this series is to provide texts which lay out the technical, environmental and political issues relating to the various proposed policies for responding to climate change. The focus is not primarily on the science of climate change, or on the technological detail, although there will be accounts of the state of the art, to aid assessment of the viability of the various options. However, the main focus is the policy conflicts over which strategy to pursue. The series adopts a critical approach and attempts to identify flaws in emerging policies, propositions and assertions. In particular, it seeks to illuminate counter-intuitive assessments, conclusions and new perspectives. The aim is not simply to map the debates, but to explore their structure, their underlying assumptions and their limitations. Texts are incisive and authoritative sources of critical analysis and commentary, clearly indicating the divergent views that have emerged and also identifying the shortcomings of these views.

This book provides a broad overview of the politics of climate change, drawing on International Relations theory and emergent concepts of climate justice. Rather than looking at the process of negotiation on international mitigation of greenhouse gas emissions in isolation, it locates it in the wider and rapidly changing international political system. Moreover, it attempts to show not only that the global climate policy regime has responded to broader power relationships in the international system, but also that part of the motivation of state participants can be usefully understood through the application of constructivist theory that emphasises the significance of identity construction and the pursuit of prestige.

Acknowledgements

Work on this book began just before the disappointment of the 2009 Copenhagen Conference of the UN Framework Convention on Climate Change (UNFCCC). Research was funded by the UK's Economic and Social Research Council via the ESRC Centre for Climate Change Economics. Within the Centre, now into its second phase, the author worked with Hannes Stephan and Robert Falkner on, amongst other things, political and institutional responses to the problems that had been highlighted by the 'failure' at Copenhagen. This raised questions of whether the apparatus of inter-state co-operation could ever make an effective contribution to slowing or stabilising the rate at which anthropogenic climate change was occurring and, increasingly, in responding to the pressing need to adapt to the damaging change that already appeared to be 'locked in' by past and current greenhouse gas emissions. Underlying this problem was an awareness that, over the life of the UNFCCC, dramatic structural change had occurred in the international system but that the links between the wider arena of inter-state politics and the specifics of climate negotiations were ill understood. This was clearly a task for the discipline of International Relations (IR) and is reflected in the title of this book: 'Climate Change in World Politics'.

My interest in the international politics of the environment and the global commons extends back to the 1980s. In thinking about these questions I benefitted greatly from discussions with colleagues in the British International Studies Association Working Group on the Environment which I had the good fortune to convene from 1992 until 2012. The imprint of these years can be seen in parts of the current book both in terms of events, such as the Rio Earth Summit and the fate of the 1982 UN Law of the Sea Convention, and also in terms of extended friendly controversies over such matters as the continuing significance of the state and the relative importance of transnational and private environmental advocacy and governance.

The School of Politics, International Relations and the Environment (SPIRE) at Keele University provided a very special setting within which to work, with its stimulating combination of green political thinkers and government and IR scholars with environmental interests. During my time as a professor at Keele we hosted a number of ECPR Green Politics Summer Schools and an EU-funded Marie Curie training site.

This brought together doctoral students from across Europe with those from Keele that I directly supervised. It has been a delight, in writing this book, to be able to cite some of their work and to observe the progress that they have achieved in their academic careers.

It would tax the reader to mention by name all the members of the groups described above who have influenced my thinking and with whom I have had the pleasure of working. I am sure that if they read this they will know who they are! There are some people I would like to name because I owe them a particular debt in terms of writing this book. Jo Lea generated the charts and diagrams for me and Lily Hamourtziadou assisted in the production of early drafts. I would also like to mention Asami Miyazaki and Duncan Weaver, both Keele postgraduate students, who used their special expertise to give a critical view of individual chapters. Radoslav Dimitrov was kind enough to read a substantial portion of the book. I am indebted to him for his invaluable perspective combining day-to-day practical diplomatic experience as a delegate with the approach of a scholar. My Keele colleague Dave Scrivener took up the burden of reading the whole book in draft and, along with an anonymous reviewer, made important suggestions for improvement. Dave possesses both an extraordinarily deep knowledge of international environmental issues, on which I have often relied, alongside a meticulous approach to the drafting of text. Finally, I owe an enormous debt of gratitude to my collaborator and partner, Charlotte Bretherton. An established scholar of the EU herself, she not only had to read, criticise and correct various drafts but also had to survive the domestic consequences of book production. It goes without saying that, while acknowledging all the inspiration and help that I have received, the errors, misinterpretations and omissions in this book are entirely my responsibility.

List of Abbreviations

AAUs	Assigned Amount Units
ADP	*Ad hoc* Working Group on the Durban Platform for Enhanced Action
AGBM	*Ad hoc* Group on the Berlin Mandate
AILAC	Association of Independent Latin American and Caribbean States
ALBA	Bolivarian Alliance of the Peoples of Our America
ANC	African National Congress
AOSIS	Association of Small Island States
APEC	Asia-Pacific Economic Cooperation
ASEAN	Association of South East Asian Nations
AWG-KP	*Ad hoc* Working Group on Further Commitments for Annex I Parties under the Kyoto Protocol
AWG-LCA	*Ad hoc* Working Group on Long-term Cooperative Action under the Convention
BASIC	Brazil, South Africa, India and China
BAU	business as usual
BOCM	Bilateral Offset Crediting Mechanism
CBDR-RC	Common but differentiated responsibilities and respective capabilities
CCS	carbon capture and storage
CDM	Clean Development Mechanism
CDR	carbon dioxide removal
CERs	Certified Emission Reduction Units
CFCs	chlorofluorocarbons
CMP	Conference of the Parties serving as the Meeting of the Parties of the Kyoto Protocol
COP	Conference of the Parties
DG	Directorate-General

ECOSOC	Economic and Social Council
EIG	Environmental Integrity Group
EIT	Economies in transition (to a market economy)
ENB	Earth Negotiations Bulletin
ENMOD	Convention on the Prohibition of Military or any other Hostile Use of Environmental Modification Techniques
EPA	Environmental Protection Agency
ERUs	Emissions Reduction Units
ETS	Emissions Trading Scheme
EU	European Union
FYROM	Former Yugoslav Republic of Macedonia
G8	Group of 8 (industrialised countries)
G20	Group of 20 (major economies)
G77	Group of 77 (developing countries)
GATT	General Agreement on Tariffs and Trade
GCF	Green Climate Fund
GCI	Global Commons Institute
GDP	gross domestic product
GEF	Global Environment Facility
GHG	greenhouse gas
GtC	gigatonnes carbon
HCFC	hydrochlorofluorocarbons
HFC	hydrofluorocarbons
HIV/AIDS	human immunodeficiency virus infection/acquired immune deficiency syndrome
IAR	International Assessment and Review
ICA	International Consultation and Assessment
ICAO	International Civil Aviation Organization
IEA	International Energy Agency
IISA	International Institute for Applied Systems Analysis
IMF	International Monetary Fund
IMO	International Maritime Organization

INC	Intergovernmental Negotiating Committee
INDCs	Intended Nationally Determined Contributions
IPCC	Intergovernmental Panel on Climate Change
IPRs	Intellectual property rights
IR	International Relations
IRENA	International Renewal Energy Agency
JI	Joint Implementation
JUSSCANZ	Japan, US, Switzerland, Canada, Australia and New Zealand
LDCs	least developed countries
LMDC	Like-Minded Group of Developing Countries
LRTAP	Long Range Transboundary Air Pollution
LULUCF	Land-use, land-use change and forestry
MARPOL	International Convention for the Prevention of Pollution from Ships
MEF	Major Economies Forum
MRV	measuring, reporting and verification
NAMAs	nationally appropriate mitigation actions
NATO	North Atlantic Treaty Organization
NGO	non-governmental organisation
NIEO	New International Economic Order
OECD	Organisation for Economic Co-operation and Development
OPEC	Organization of Petroleum Exporting Countries
P5	Permanent members of the United Nations Security Council
ppm	parts per million by volume
PRC	People's Republic of China
PwC	PricewaterhouseCoopers
QELRCs	quantified emissions limitation or reduction commitments
QELROs	quantified emissions limitation or reduction objectives
REDD+	reducing emissions from deforestation and degradation in developing countries, including conservation

REIO	Regional Economic Integration Organisation
RIIA	Royal Institute of International Affairs
SBI	Subsidiary Body for Implementation
SBSTA	Subsidiary Body for Scientific and Technological Advice
SIDS	Small Island Developing States
SRM	solar radiation management
TEC	Technology Executive Committee
TFEU	Treaty on the Functioning of the European Union
UAE	United Arab Emirates
UK	United Kingdom
UN	United Nations
UNCHE	United Nations Conference on the Human Environment
UNCLOS	United Nations Convention on the Law of the Sea
UNCTAD	United Nations Conference on Trade and Development
UNDP	United Nations Development Programme
UNEP	United Nations Environment Programme
UNFCCC	United Nations Framework Convention on Climate Change
UNGA	United Nations General Assembly
US	United States
WCED	World Commission on Environment and Development
WRI	World Resources Institute
WTO	World Trade Organization

1
Introduction

There are already a great many books about climate change and some very good ones about its international dimensions. Any author venturing into this crowded field needs to provide a justification. Here is mine. It is essentially twofold. First, that the importance of international politics, in the sense of relations between sovereign states rather than the transnational and non-state phenomena that now occupy so much attention in academic studies of the international relations of global environmental change, deserves if not re-instatement then certainly a re-statement. Second, those studies of international environmental cooperation, now commonly described as 'global governance', have become rather divorced from the world political context that surrounds them. This might not matter so much for functional negotiations on highly technical aspects of transborder pollution, but it will be significant for the long-running attempt to create a comprehensive and effective international climate regime. From the outset this has been widely, but not universally, recognised as something of critical importance for planetary survival and has been accorded a political status which marks it out from more mundane environmental issue areas. Conferences of the Parties to the United Nations Framework Convention on Climate Change (UNFCCC) have been attended at the highest political level by presidents and prime ministers. United Nations (UN) Secretary Generals have summoned them to take action and climate change issues have appeared on the agenda of the Security Council, Group of 8 (G8) and, indeed, most other leading international organisations. Furthermore, the international climate regime has been constructed during 20 years of the most profound change in the international system, from which it cannot have remained isolated.

It is very understandable that much writing and research on the international relations of global environmental change has avoided a 'state-centric' approach. The interstate climate regime has moved at a glacial pace since the signature of the Kyoto Protocol in 1997. There have been numerous disappointments and frustrations as the Copenhagen Conference of the Parties (COP) of 2009 and other meetings failed to match expectations. In fact, in ways which are considered in Chapters 3 and 8 of this book, the climate regime has not only been becalmed but in some respects has moved backwards. In sharp contrast, and frequently as the direct result of perceived 'deadlock' in international negotiations, there has been a flowering of non-state, sub-state and transnational activity. This has occurred with the activities of cities (Bulkeley and Schroeder, 2012) and in the corporate sector (Clapp, 1998) to the extent that private governance is now a very important component of the academic study of global environmental politics (Pattberg, 2007).

One of the first books to address the international relations of the environment, published at the time of the signature of the UN climate change Convention, formulated the problem as follows:

> Can a fragmented and often highly conflictual political system made up of over 170 sovereign states and numerous other actors achieve the high (and historically unprecedented) levels of co-operation and policy co-ordination needed to manage environmental problems on a global scale?
>
> (Hurrell and Kingsbury, 1992, p. 1)

After 20 years of experience with the climate regime many analysts would be tempted to answer in the negative. Disillusionment with interstate cooperation goes deeper than the specific failure to produce a new and comprehensive climate agreement. It should be seen in the context of a more general concern about the continuing viability of state-based political forms. The final test of the climate regime will be its effectiveness in providing a means to manage the global atmospheric commons. Commentators have generally been dismissive of what has been achieved so far. Treaties have been made that 'are easy to agree on' yet 'had almost no impact on the emissions that cause global warming' (Victor, 2011, p. 3). The question of whether this is true, and likely to continue to be so, forms the substance of the concluding chapter of this book.

Green thinkers and radical ecologists have, in fact, identified the state itself as part of the global environmental problem. In these circumstances, international cooperation between state authorities could be a potentially damaging distraction.[1] The question of the desirability and relevance of international environmental cooperation has thus been part of a broader political and philosophical debate concerning the possibility of a 'green state' (Eckersley, 2004). A distrust of existing forms of state and government runs deep in green politics and, at the international level, was coupled with an enthusiasm for an emergent 'global civil society' represented by the very large number of environmental and development non-governmental organisations (NGOs) working on climate-related issues and appearing in force at the annual UNFCCC COPs. The failure of state governments to take decisive action and the apparent weakness of interstate processes led to a search for political alternatives in novel forms of discursive democracy and 'networked governance' (Stevenson and Dryzek, 2014).

In the light of all this, what remains for international cooperation? The considered conclusion of over 30 researchers working within the 'Earth Systems Governance Project' addresses this question:

> New governance mechanisms cannot take away from the urgent need for effective and decisive governmental action, both at the national and inter-governmental level. Governance beyond the nation state can sometimes be a useful supplement especially when they avoid being captured by powerful interests and instead focus on problem amelioration. Yet even for this, it requires support and oversight from national governments.
>
> (Biermann et al., 2012, p. 5)

In an attempt to defend the practice of international environmental cooperation between governments, I have argued that there are certain functions that need to be performed by nation-states as presently constituted, at least within any time frame that is relevant to dealing with the climate change problem (Vogler, 2005). Establishing what these may be is important because it can provide a perspective on what may reasonably be expected of international cooperation. All too often a lack of clarity about the, sometimes limited, contribution that can be made to the solution of problems at the international level can lead to disillusionment and a rejection of the entire process. The point has frequently been made that problems that are conceptualised on a global scale do not necessarily require fully global solutions. All that may be required is an orchestration of local and regional actions. Norm creation and propagation is one such orchestrating function with which the

UN system, since 1972, has been involved. A famous example, written in the 1992 Climate Convention and extensively referenced elsewhere is the principle of 'common but differentiated responsibilities and respective capabilities' (CBDR-RC).

Chapter 2 of this book takes up the issue of how prevalent international norms and understandings 'frame' the issue of climate change, defining both substance and the limits of the possible. Much effort has been expended on encouraging private sector funding for climate mitigation and adaptation, but it remains the case that it is only state authorities that are in a position to mobilise the resources required to build capacity among less-developed countries and to provide aid and assistance. Both are highly significant elements in the operation and politics of the climate regime. The same point can be made for information-gathering and scientific work. There is really no alternative to international cooperation and funding of bodies such as the, significantly named, Intergovernmental Panel on Climate Change (IPCC). The private sector may undertake pharmaceutical and other types of commercially relevant research but it will lack the incentives and authority to engage in basic climate science and the compilation of inventory data.

To this list of state functions in the international environmental realm one might normally add the regulation of transboundary flows of pollution and goods, and there are parts of the global climate problem for which this is relevant activity. However, the fundamental requirement for interstate action arises from an understanding that the atmosphere constitutes one of the global commons. Because it is beyond sovereign jurisdiction there are incentives that drive excessive exploitation in terms of anthropogenic emissions of greenhouse gases (GHGs) with the 'tragic' result of dangerous climate change. State authorities need to impose some regulatory control to mitigate these emissions, in the same way as other commons are governed through voluntary action by users. The critical issue, reflecting the competition and distrust that exist between users, is to ensure that 'free riding' does not occur. Users of the commons will need to be assured that any efforts that they make to reduce their polluting emissions will not be exploited by others who fail to make equivalent reductions. This is, in essence, the economist's view of climate change as a collective action problem where climate change represents the world's greatest market failure in the provision of the ultimate public good. Here, the role of international cooperation is expressed in terms of action to: '... overcome the market failures that lead to the under-provision of public goods where individuals or countries face an incentive to free ride on the actions of others' (Stern, 2007, p. 45).

In the global system only cooperating governments are in a position to agree and impose such controls, and this remains the central functional requirement of an effective climate regime. Radical critics sometimes portray the state as being trapped within the global structure of capitalist accumulation and incapable of independent agency. This is unduly fatalistic, for there is evidence to suggest that governments have on occasion summoned up the will to make the changes required for, to quote a famous example, the restitution of the stratospheric ozone layer (in fact one of the problems with the climate regime was that, as will be argued in Chapter 2, it was overly influenced by the success of the Montreal Protocol on Substances that Deplete the Ozone Layer).

It also continues to be the case that nation-states remain the focus of loyalty and are, in the absence of a central world government, the only agents possessing sufficient capability and legitimacy to orchestrate the regulatory action necessary to sustain the global atmospheric common. A serious qualification needs to be added concerning the undifferentiated use of the category 'nation state'. In reality we are dealing with a class of state actors which do have effective governments and control of resources, to which we may add the European Union (EU), when acting within its climate-related competences. A substantial number of the state Parties to the UNFCCC would not fulfil these requirements. They are often miserably poor, highly vulnerable to the impacts of climate change and lacking in resources, effective internal government and the capability to engage in anything more than minimal participation in the climate regime. It is for this reason that the provision of aid and capacity building represents such a significant element of the climate regime.

However, it is evident that a great deal of the activity to be observed in international climate politics does not necessarily accord with the stated purposes of the regime. The pursuit of very specific national interests, often determined by questions of competitiveness and energy security, will be evident alongside regional conflicts and the politics of organisational status within the UN system. The UNFCCC should be conceived of as one arena among many in a long-term North–South confrontation over economic development and environmental responsibility and justice. It is also infected by struggles for recognition and prestige which have always interacted with the dynamics of power relations between states. In the background are the momentous structural changes that have transformed the global economy and international political system since the beginnings of the climate regime in the 1980s. Such things are, or perhaps ought to be, the stock-in-trade of the academic study of International Relations (IR).

Climate change and environmental issues in general have often sat uneasily within the discipline of IR. There are several reasons for this. Climate change was for a long time seen as a rather specialised area, dependent on an understanding of a contested and difficult science and negotiated by technical experts operating within an arcane and complicated regime. The overwhelming bulk of scholarship was performed within a rational choice and liberal-institutionalist paradigm that took as its main problematic the solution of collective action problems. This has been observed by outside critics for some time (Smith, 1993). There is no space here to review the extensive literature on international environmental politics, a task that has been ably performed elsewhere (O'Neil, K., 2009), but institutionalist approaches are still prominent alongside studies of NGOs and transnational activity. Their shared, and very understandable, preoccupation is with governance. The difficulty is that, as I have argued elsewhere, this is often to the exclusion of politics (Vogler, 2012). Approaches to IR that are prominent in the discipline, including realist power politics, normative analysis, English School reflection on the nature of international society and constructivist studies of the politics of identity, have not been very evident in the climate change literature. In consequence, it may be argued that the international climate regime tends to be treated endogenously, in both empirical and theoretical isolation. Much of the literature on international environmental cooperation can also be characterised as having a functionalist orientation towards the conclusion of effective international agreements. The functional approach to IR has a long history that refers back to the great public international unions of the nineteenth century and to later schemes by David Mitrany (1975) and others to circumvent the sovereign sensitivities of statesmen, and their explicitly political differences, by organising low-level cooperation across national borders for the solution of shared economic, social and welfare problems. A review of work by Oran Young (2010), who has been the leading theorist in the field, makes the point explicitly:

> One of the greatest challenges to improving our understanding of global environmental governance is acknowledging the excessive functionalism of much recent research ... It is entirely possible that institutions are also created for functionalist purposes – but it is not axiomatic. Institutions may also be functional for states precisely

because they are weak. Politicians may find value in supporting institutions that provide little more than symbolic benefit ... The strong functionalism implicit in many strands of research on environmental governance renders them unable to make sense of these dynamics. It also leads them to systematically underestimate the political obstacles facing some environmental regimes.

(Marcoux, 2011, p. 147)

This book attempts to take up this challenge by conducting a political investigation of the ways in which the international community has sought to deal with the complex and difficult problem of climate change. It asks questions about how and why the climate problem has been framed in a particular and fragmentary way, leading to responses that appear to neglect some of the key socioeconomic drivers of the enhanced greenhouse effect. It goes on to consider the motivations and national interests of the state Parties to the UNFCCC and the alliances that dominate the politics of that institution. Part of the explanation of why it has proved so difficult to arrive at a comprehensive post-Kyoto climate agreement, is to be found in the incompatibility of perceived national economic interests and the disconnection between national responsibility and vulnerability to effects of alterations in the climate. But this is by no means the whole story. There is also the indissoluble relationship between the climate regime and the demands for restitution and fairness that motivates developing countries, leading to the issue of what exactly 'climate justice' means at the international level and whether it is separable from the pursuit of material national interests. In common with many areas of international life, symbolic politics is an evident dimension of international climate discussions, and it will be amplified when the climate is linked to security issues or discussed at the level of heads of state or government. There is a need to consider who benefits from prestige- and recognition-seeking activities and what they may mean for the possibilities of agreement.

Underlying any international political analysis are questions of power in both its relational and structural forms. How is power exercised within the climate regime and to what effect? Questions of national interest and motivation involve agency, but agency is constrained and conditioned by structures. Much of the research on the climate regime tends to be focused on the regime itself, rather than the wider structural context of the international system. A further, and very difficult, question concerns the extent to which the fate of the regime, and power relations within it, are determined by the overall international economic and political structures. Most significant in all of this is the shifting relationship between the climate regime and trends in the wider international system within which it is embedded. At the primary level of analysis this denotes the international system of states, but beneath this are the shifting and crisis-ridden structures of the global economy which have driven both the exponential rise in GHGs and the alteration in interstate power balances. This is especially important in an era when these structures have been subject to very substantial changes.

In grappling with these questions I have resorted to various types of IR theory, where appropriate and where they provide tools of analysis for the dissection of national interest or the influence of structure. This may be criticised as theoretical eclecticism, but I found it useful to think through the classic dialogues between liberalism and realism or between communitarianism and cosmopolitanism in relation to the specifics of the climate regime. In Chapter 2, I have used the framework of regime analysis to organise discussion of the way the climate 'issue area' was defined, and in Chapter 3 the description of the UNFCCC follows established practice in terms of its outline of principles, norms, rules and decision-making procedures. The influence of constructivism is evident in the discussion of 'framing and fragmentation' in Chapter 2 and again, to a lesser extent, in the discussion of EU identity creation in Chapter 6.[2]

An outline of the book

The book is organised as follows. It commences with a discussion of the way in which the climate regime was constructed, followed by an analysis of its key characteristics and changes up until the beginning of 2015. There then follow three chapters that cover agency in international climate politics, starting with national interests and then moving to the related questions of the pursuit of justice and the search for recognition and prestige. Structural change and explanation are covered in Chapter 7. The concluding chapter reflects on the effectiveness of the climate regime and some implications of the political obstacles and opportunities identified in the preceding chapters.

Chapter 2: Framing and fragmentation

Issue areas are politically derived constructs which often define a problem in a partial way. The inclusions and exclusions in the UNFCCC and its Kyoto Protocol are considered in specific terms, for example the concentration on territorial emissions and the exclusion of emissions arising from international aviation and sea transport. Broader questions are also raised as to why, what might be considered by independent observers as essential drivers of anthropogenic emissions, such as increases in population and consumption, do not appear on the agenda. On the other hand, there have been serious attempts to redefine the climate problem in ways which link emissions mitigation and adaptation to security. The absences from, and fragmentation of, what is commonly described as the climate 'regime complex' may be explained not only by the way in which the issues are framed but by the self-interested manoeuvres of governments and organisational bureaucracies.

Chapter 3: The UNFCCC regime

Here, the central international climate regime is analysed as a global commons regime. The evolution of its principles and norms is reviewed. The principle of common but differentiated responsibilities, which has exempted Group of 77 (G77) members from making mandatory emission reductions, has come under increasing strain as developing countries are predicted to become responsible for the majority of global emissions by the 2020s. The Kyoto Protocol represented one approach to organising the mitigation activities of developed countries. It has largely been replaced by a looser but comprehensive and differentiated system of 'bottom up' voluntary 'contributions'. The regime has evolved complex rules for information gathering and for the operation of the flexibility mechanisms of the Kyoto Protocol. There has been slow but evident progress in areas such as adaptation funding and technology transfer. The history of the UNFCCC has been punctuated by attempts to use development funding as a lever to promote G77 adherence to the regime – as with current adaptation measures and development of arrangements for 'reducing emissions from deforestation and degradation in developing countries, including conservation' (REDD+). The internal political system of the regime is also considered, with its proliferation of sub groups and committees and the continuing failure to agree on voting procedures, which have provided ample scope for obstruction.

Chapter 4: Interests and alignments

This chapter employs an orthodox definition of national interests to set out the varying motives of the Parties to the Convention. Unlike many other environmental regimes, the interests engaged are unlikely to fall within the category of 'milieu goals' for general improvement. Instead they are 'possession goals' which are likely to be taken most seriously by governments. Such goals are seen as a balance between national economic and energy security interests and perceptions of cost and vulnerability to the impacts of climate change. Climate politics can be portrayed as a nexus of the environmental agenda of the North and the developmental demands of the South, and it is important to relate what has occurred within the regime to broader changes in North–South relations. Developing countries have very different stakes in the climate negotiations, but what unites them is the concept of the primacy of development over GHG mitigation and adherence to CBDR-RC principles. The chapter is structured around the major negotiating groups: The European Union, The Umbrella Group, Environmental Integrity Group and G77. Within the latter are the Alliance of Small Island States (AOSIS), the members of the Organisation of Petroleum Exporting Countries (OPEC), the Less Developed and, since 2009, Brazil, South Africa, India and China, the BASIC group of large emerging economies. Special attention is paid to positions adopted on a 2015 agreement and recent crosscutting alignments, the Cartagena Dialogue and the Like-Minded Developing Countries. The benefits, along with the costs of climate change, are unevenly distributed in other ways and a simple North–South dichotomy is no longer adequate to describe the cleavages between rich and poor and centre and periphery in both hemispheres.

Chapter 5: The pursuit of justice

Only a narrow realism would deny the significance of normative considerations alongside national interests. Attempts to mitigate emissions and adapt to climate change are essentially inseparable from questions of climate justice and the obligations of those primarily responsible for the emissions that cause climate change to those who are vulnerable and/or less developed. The question of distributive climate justice, as between developed and developing worlds, has never been resolved and continues to provide a basis for asserting national rights and claiming legitimacy. This chapter employs the distinction between communitarian and cosmopolitan conceptions of international ethics to investigate the nature of the different proposals on fairness, the distribution of mitigation commitments in relation to current and cumulative emissions and occupation of the 'carbon space'. This helps to not only clarify the differences between the contending nations, but also reveals strong links to national economic interests and the failure to consider a truly cosmopolitan approach that would establish the rights of individual human beings. Nonetheless, the climate regime demonstrates the beginnings of a communitarian approach through developing principles of assistance with adaptation and compensation for loss and damage.

Chapter 6: Recognition and prestige

The starting point of this chapter is the recent revival in IR theory of the study of identity and recognition as elements of state policy. The symbolic politics that arise from this can account for some of the significant and sometimes apparently irrational behaviour that occurs within the climate regime, which is often neglected in rationalistic interest-based accounts of negotiations but is readily apparent to participants. Climate diplomacy resonates with assertions of sovereignty and national prestige because, it will be argued, it has diverged from 'normal' international environmental negotiations, attaining a much higher political profile. The final week of the 2009 Copenhagen COP provides the most graphic illustration of this trend. Three examples are used to illustrate these propositions. Japan, denied some of the normal trappings of great power status, attempted to use sponsorship of environmental conventions as a means of improving its international standing. The climate 'leadership' role of the EU is analysed as a case study in identity construction in which symbolic opposition to the United States (US) became an important justification of the Union as an international actor. Then there is the very different case of the members of ALBA (the Bolivarian Alliance of the Peoples of Our America), who used dissent from the agreement reached by the big climate powers at Copenhagen to register their independence and opposition to 'imperialism'. Finally, this chapter considers the status conferred by the membership of select international 'clubs' such as the G8, G20 and the Major Economies Forum.

Chapter 7: Structural change and climate politics

The preceding chapters focus on interpretations of agency. The purpose of this chapter is to set these actions in their structural context. A fundamental thesis of the book is that the climate change regime, which has become very different from a 'normal' or 'functional' type of international environmental cooperation, should not be viewed in isolation. This chapter attempts to consider in detail the extent to which changes in the climate regime can be associated with the dramatic shifts occurring in its wider political environment, including the collapse of bipolarity and the changed structure that emerged after 2000. Underlying economic trends in the system are considered in relation to the way in which they both contributed to increases in emissions of GHGs and shaped the international political structure. Structural interpretations matter because they can provide explanations of power relationships and the determination of outcomes. Clearly the overall 'hegemony' of the United States and the subsequent rise of the BASIC countries provide significant structural influences over the course of the climate regime. However, the specific structural relationships within the climate regime itself, as part of a UN General Assembly-based system, have provided a countervailing source of power.

Chapter 8: Conclusion

In the end, questions of the effectiveness of the regime and its future development cannot and should not be avoided. This concluding chapter reviews evidence on its impact on governmental and corporate behaviour in relation to reducing emissions to achieve stabilisation at or below the 2 °C threshold. Current trends towards formulating a new agreement to be operative in 2020 are considered in the light of the political evolution of the regime. The fact that the UNFCCC cannot be extracted from its politico-economic context limits the scope for purely institutional solutions of the type that may be applicable to other, less contentious, environmental regimes. The climate regime may also be incapable of accommodating those demands for climate justice that are not necessarily represented by the 'state Parties'. This leads to analysis of what role, if any, the UNFCCC can be expected to play. The political dimensions of the regime that are examined in this book certainly complicate and obstruct the search for agreement, but there may also be ways in which they can be utilised to provide a more positive outcome.

2
Framing and Fragmentation

> Global climate change is the apotheosis of the idea that everything is related to everything else.
> (Eugene Skolnikoff, 1993, p. 183)

Even a cursory glance at international attempts to solve the problem of climate change would suffice to establish two things. First, that, although there is a UN Climate Convention (UNFCCC) and associated Kyoto Protocol, there is also a plethora of other climate-related initiatives and institutions. Second, there is a significant disconnection between the way in which the UNFCCC attempts to mitigate climate change and scientific and even 'common sense' ideas of what would really be required to tackle the human forces that drive the enhanced greenhouse effect. This chapter examines the way in which the international climate regime was set up and how 'framings' of the problem have been associated with the 'fragmented' responses of the international community.

The concept of framing has been an important one across the social sciences. The discussion that follows is based on some social constructivist assumptions that utilise the insights of the philosopher John Searle (1995). In Searle's conception, physical climate change is an 'observer independent' set of 'brute facts'. We may comprehend and 'socially construct' such facts imperfectly, at best through intersubjective agreement between natural scientists, but also through our own direct physical experiences. Erving Goffmann, the sociologist who first coined the idea of 'frame analysis', referred to the ways in which humans build up their definition of a situation 'in accordance with principles of organization which govern events – at least social ones – and our subjective involvement in them' (Goffmann, 1986). For Goffmann there are two types of primary framework; natural and social. Natural frameworks are purely physical, with natural determinants. 'Elegant *versions* (my italics) of these natural frameworks are found, of course, in the physical and biological sciences' (ibid., p. 22). Social frameworks, by contrast, provide background understanding of events 'that incorporate the will, aim, and controlling effort of an intelligence, a live agency, the chief one being a human being' (ibid.). For a social constructivist, such as Searle, both the elegant scientific versions of reality and the various other socially constructed frames found in academia, policymaking and diplomacy are all 'social facts', as are the institutions of global governance.[1]

Adopting this perspective means that an enquiry into the relationship between various climate-related frames and institutional structures is, in effect, a study of the connections between various levels and types of social facts. The frequently posed question of why there is such a disjunction between a, largely shared, scientific version of the likely effects and drivers of climate change, and the policy frames and agendas adopted by governments, reflects an implicit ranking. Scientific versions of reality are seen as superior to policy frames and have not been adequately translated to the latter. The former often represent a holistic vision of the earth's physical and social systems in which there are myriad interconnections, rendering climate change an infinitely more complex and 'wicked' problem in which 'everything is related to everything else'. Very often comparisons are made with the problem of stratospheric ozone depletion, where both scientific and policymaking frames were able to converge on a simple set of variables. Much recent literature on policy formation has been concerned with the discursive frames that set the terms of possible argument, serving to delimit and decompose problems (Scrase and Ockwell, 2010).

Because of the inherent complexity of the climate change problem such decomposition is especially important. The broadest frames are coincident with the principles of the prevailing international order and would often tend to favour solutions based on free markets and private sector involvement. Alternatives that might once have figured, for example the institution of common heritage of humankind principles, are, after the end of the Cold War and the associated dominance of liberal economic ideas, beyond consideration. There are other key inclusions and exclusions which will be considered below. A framing of the climate problem in terms of security has certainly gained traction at the international level in the past decade but others, such as consumption and population increase, which figure in the scientific and more popular literature on the drivers of climate change, are notably absent.

The key link between policy frames and the international institutional architecture is to be found in the way that they prefigure the setting of policy agendas involving the selection of issues. In the international cooperation literature, regimes govern specific 'issue areas'. The latter are defined as 'sets of issues that are in fact dealt with in common negotiations and by the same and closely coordinated bureaucracies' (Keohane, 1984, p. 61). This is a largely intergovernmental activity, although the non-governmental organisations and interest groups can exert significant influence. The Brundtland Commission's Report, that preceded the 1992 Rio Earth Summit, recognised that the internal structure of governments could have a major influence on the framing of issues – the separation of trade and environment for example (World Commission on Environment and Development [WCED], 1987, p. 310). Both this and the need to accommodate to the architecture of international organisations, with its existing division of responsibilities and budgets, seem to have been influential in the way in which the climate issue area was defined. The critical point, in terms of the likely effectiveness of regime arrangements, is that such definitions and delimitations are often inconsistent with other framings – notably those of natural science (Vogler, 2000, p. 24).

It is difficult to find examples of international regimes that address problems in an encompassing and coherent way. Usually they are partial and fragmented. According to Biermann et al. (2009, p. 31) such fragmentation is a 'ubiquitous structural characteristic of global governance architectures today'. The most notable exception to this rule is the nearly all-encompassing set of legal rules for the definition and usage of the oceans contained in the 1982 UN Convention on the Law of the Sea (that entered into force in 1996 after substantial revision of its Part XI on the deep sea bed). The climate 'regime complex', with the Convention (UNFCCC) at its heart, provides a case study in fragmentation, where the efforts of governments have 'produced a complex of more or less closely connected regulatory regimes' (Keohane and Victor, 2010, p. 1).There is some fragmentation even within the formal structures of the UNFCCC, with its extensive series of meetings, negotiation tracks and subsidiary bodies. Beyond it there exists a very large number of public and private organisations that would merit inclusion, such that their description would require a complex mapping exercise. Abbott (2012, p. 571) refers to a 'cambrian explosion' of 'transnational institutions, standards and financing arrangements', the overall climate regime being 'complex, fragmented and decentralised'.

The focus of fragmentation studies usually reflects the 'problem - solving' frame of liberal institutionalist IR. The questions that are asked often relate to the degree to which fragmentation helps or hinders the search for cooperative solutions to global governance problems in general, and mitigation and adaptation to climate change in particular. The task of this chapter is rather more modest. It is primarily to consider how framing ideas and the fragmentation of the international climate architecture fit together. In the first part, specific inclusions and omissions within the UNFCCC will be considered and, in the second, some of the broader framings of the climate problem.

Framing the climate convention and Kyoto Protocol

The 1972 United Nations Conference on the Human Environment (UNCHE) is often regarded as the event that marked the emergence of environmental issues on the international agenda. There was no mention of climate change in the 'principles' drafted by the conference but it was noted in one of 18 accompanying recommendations. This proposed the establishment of a world wide network '... to monitor long term trends in the atmosphere which might cause changes in meteorological properties, including climatic changes' (UN, 1975, p. 322). Quasi-official preparatory work for the conference, involving consultations with 152 prominent experts from 58 countries, had clearly recognised that, despite prevailing scientific uncertainties, there was a real climate problem that the international community would have to face. The words of the preparatory commission are worth quoting:

> But the balance between incoming and outgoing radiation, the interplay of forces which preserve the average global level of temperature, appear to be so even, so precise that only the slightest shift in the energy balance could disrupt the system. It takes only the smallest movement at its fulcrum to swing a seesaw out of the horizontal. It may require only a small percentage of change in the planet's balance of energy to modify temperature by 2˚C.
> (Ward and Dubos, 1972, p. 266)

The reference to 2 °C may appear prescient, but it should also be recalled that, although an estimated 0.5 °C rise in global mean temperatures was predicted by 2000, it was acknowledged that there could equally well be a movement in the opposite direction – towards 'global cooling'. At the end of the 1970s such scientific agnosticism was still present in a survey of the best available scientific opinion that assigned equal probability scores to global cooling and warming.[2]

The initial framing of the climate problem in the 1992 UNFCCC, representing the achievable consensus of the late 1980s, has remained of central importance. First, climate change, in the first two lines of the Convention, is acknowledged as a 'common concern'. That is to say the atmosphere is not to be treated as a 'common heritage', with all the legal properties of shared ownership and enjoyment that might be implied on the basis of experience with the designation of deep seabed mineral resources (and the moon) as the 'common heritage of humankind'. Malta, which had initiated the seabed version of common heritage in 1967, made the suggestion that the climate be treated as part of the common heritage at the first discussion of climate issues at the UN General Assembly in 1988 (Bodansky, 1993, p. 465). This did not prove to be acceptable and the final United Nations General Assembly (UNGA) resolution refers to the climate as a common 'concern' (UNGA Res. 43/53, 1988).

The influence of experience with negotiations for the UN Convention on the Law of the Sea (UNCLOS), which had controversially adopted the idea of common heritage, was also evident in the format of the projected climate agreement. The Conference had run on from 1973 to 1982 and concluded with a vast and comprehensive treaty covering almost the entire range of maritime issues. At the time of the negotiation of the UNFCCC it was still awaiting sufficient ratifications for entry into force. The desire to avoid a comparable experience with the international regulation of atmospheric issues probably assisted the rejection of Canadian proposals for a broad and comprehensive 'law of the atmosphere' that would cover a range of interdependent atmospheric issues, including climate change (Bodansky, 1993, p. 472). Instead, the clear preference was for a narrower framework agreement, along the lines of the 1985 Vienna Convention on substances that deplete the stratospheric ozone layer. A widely regarded 'control' Protocol to this Convention had been adopted in 1987. The Montreal Protocol was specifically designed to ban production and trade in those chlorofluorocarbons identified as having stratospheric ozone-depleting effects. Instead of following this example, the climate regime was to target a range of ubiquitous 'source' GHGs and 'sinks'[3] when it might, for example, have been more productive to arrange the reduction or phasing out of coal production and trade. However, the difficulties of doing this and arranging compensation for economies dependent on coal would have been immense.

If the major part of the climate problem arises from the burning of hydrocarbons in energy generation (constituting some 80 per cent of global energy supply) then it would also have made sense to target government fuel subsidies and to institute measures to encourage the development of renewables. David Victor has argued that the difficulty with international attempts to manage the climate problem is that it was initially framed as an environmental issue whereas '... in reality, its root causes and solutions lie in the functioning of energy markets and in the incentives for technological change within those markets' (Victor, 2011, p. xix). National mitigation actions envisaged under the Convention were at some remove from the direct sources of the problem. Hence, the efforts of governments to fulfil their emissions reductions obligations under the Kyoto Protocol, and a successor agreement, have been beset by factors such as the extreme gyrations of the global oil price. In the EU, for example, long-range plans to decarbonise and set ambitious GHG reduction targets have been destabilised by energy price movements that undercut the assumptions of renewables policy and weaken its Emissions Trading Scheme (ETS) (Keating, 2015).

The relationship between energy and climate is of primary significance, but it is a connection imperfectly made by policymakers and in the international institutional architecture of the International Energy Agency (IEA) and other bodies. The regime complex for global energy governance, insofar as it exists, is both fragmented and fundamentally oriented towards security of supply rather than sustainability. In a companion study to this volume, Thijs Van de Graaf (2013) demonstrates how the failure to provide effective management of global energy has arisen from its inherent functional characteristics but also as a result of strategic behaviour and organisational rigidities. While the IEA has begun to re-orientate its activities towards the energy-climate nexus, 'both spheres are still largely disconnected' (ibid., 102). Thus international action in the promotion of renewables remains in its infancy with the creation, in 2009, of the International Renewable Energy Agency (IRENA). An integrated approach to climate and energy issues has only received spasmodic attention from the G8 and G20 meetings, the former having a long-term interest in energy security, complemented from 2005 by the assumption of climate change responsibilities. There have been many bilateral and some multilateral attempts to foster international technological cooperation in the development of renewable alternatives to fossil fuels, cleaner coal-fired power generation or carbon capture and storage but, as with the Asia Pacific Economic Cooperation (APEC) countries' climate initiative, led by the Bush administration, these have sometimes been presented as alternative, rather than complementary, to mitigation under the UNFCCC Kyoto Protocol.

The Climate Convention was initiated and negotiated under the auspices of the UN General Assembly and in anticipation of the 1992 Earth Summit. It was, accordingly, framed in terms of developmental politics and the obligations of the North to the 'Global South'. Emblematic of this was the famous, or in some quarters infamous, principle of CBDR-RC. This assigns the initial responsibility for making emissions reductions to the developed (Annex I) countries and is carried through in the requirements of the 1997 Kyoto Protocol, under which the developed country Parties were to achieve a collective 5.2 per cent reduction against a 1990 baseline. The interpretation and continuing relevance of CBDR-RC have been a dominant theme in the development of the UNFCCC and is the subject of Chapter 5 of this book, concerning the politics of climate justice. The 'respective capabilities' part of this principle was often neglected, but it foreshadowed a time when the emergent economies of the South would also have to engage with emissions reductions.

GHG mitigation under the Convention treats both sources and sinks on a *territorial* basis. It is the responsibility of state governments to fulfil their reduction and reporting obligations within their own territories. Under the Kyoto Protocol a 'basket' of six gases (CO_2, CH_4, NO_2, HFCs, PFCs and SF_6) were subject to emissions reductions and removals. Annex I inventories were to include emissions of these gases from six sectors (energy, industrial processes, solvents, agriculture, 'land-use, land-use change and forestry' [LULUCF] and waste). In principle, this might appear to include a substantial proportion of anthropogenic GHG emissions arising from power generation and industrial processes. But there were some significant omissions.

One appears straightforward – the chlorofluorocarbons (CFCs) and other ozone-layer-depleting GHGs controlled under the Montreal Protocol. The Protocol has, in fact, been extremely successful in removing ozone-depleting substances – mainly CFCs and hydrochlorofluorocarbons (HCFCs). The effect of these measures is not only measured in terms of the stratospheric ozone layer but also in the avoidance of GHG emissions that would otherwise have occurred – some 10gt of CO_2 equivalent per annum (United Nations Environment Programme [UNEP], 2011, p. 21). Unfortunately, the CFCs and HCFCs controlled under Montreal are being increasingly replaced by an 'ozone-safe' but highly climate-damaging class of chemicals – the hydrofluorocarbons (HFCs). Left unchecked their emission could undo all the GHG emissions savings achieved by the Montreal Protocol producing, by 2050, global warming effects roughly equivalent to those of global transport emissions (UNEP, 2013).[4]

Global transport emissions from international shipping and aviation also constitute a significant lacuna in the climate regime. Their GHG emissions (from aviation and marine fuel bunkers) were not excluded from consideration under the Convention but, in Article 2(2) of the Kyoto Protocol, Annex I countries were enjoined to pursue their reduction through the International Maritime Organisation (IMO) and International Civil Aviation Organisation (ICAO). In favour of this approach was the difficulty of calculating and reducing such emissions on a national basis and the long-standing international agreement that aviation fuel would not be subject to taxation. The contribution of international transport to global GHG emissions is only around 5 per cent in total but this share is rising fast and aviation emissions impose additional burdens on the climate estimated to be 2–4 per cent greater than CO_2 emissions alone (Ribeiro et al., 2007, p. 376).

While carrying around 90 per cent of world trade, the international shipping sector is responsible for around 2.7 per cent of global GHG emissions (International Chamber of Shipping, 2012, pp. 2–3). There is a long-established regulatory framework under the International Convention for the Prevention of Pollution from Ships (MARPOL), operated through the IMO, which did not, until 2012, respond to the call to mitigate the industry's GHG emissions.[5] Aviation is a much more fuel inefficient way of transporting a tonne of cargo (ships = 15 grammes per tonne/km, air freight = 540 grammes) (ibid., p. 5). It

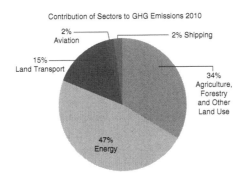

Figure 2.1 Contribution of sectors to GHG emissions 2010

Source: IPCC (2014), 'Climate Change 2014'. Available at: http://www.ipcc.ch/report/ar5/wg3/. Last Accessed: 11/07/2014.

is also the fastest rising source of transport-related GHG emissions (Figure 2.1). IPCC estimates that aviation emissions will rise two-and-one-half times between 2002 and 2030 (ibid.). Domestic aviation emissions are included in national emissions targets under Kyoto, but international emissions are completely unregulated.[6] Frustrated by the continuing lack of progress on GHG emissions by the industry and governments in the ICAO, the EU agreed, in 2008, to include emissions from international flights in its emissions trading system. The resulting conflict between the EU and other countries over the legality of this policy, the decision by the US Senate to forbid US carriers' participation and Chinese threats to retaliate against Airbus, provide an excellent illustration of the difficulties and sovereign sensibilities involved in any attempt to bring international emissions under control. In 2012 the ICAO was finally moved to consider developing its own scheme for GHG reduction.[7] There is an additional problem for aviation regulation arising from the fact that numerous developing world economies are dependent on a tourist trade that might be damaged by additional fuel costs – many of the countries involved are also small island states.

All this may lead to the conclusion that the conventional 'framing' of the climate problem is simply outmoded. Its territorial approach to emissions inevitably reflects the assumptions of the sovereign states system, while governments have every short-run incentive, as illustrated in the struggle over aviation emissions, to protect national interests that are served by current rules and assumptions.

The securitisation of the climate

There have been concerted attempts to re-frame climate change as an international security issue. This may be regarded as a 'securitisation' process involving '... a move that takes politics beyond the established rules of the game and frames the issue either as a special kind of politics or as above politics' (Buzan et al., 1998, p. 23). There was some basis for a new, 'environmentally tinged' social construction of security. Since the ending of the Cold War, and parallel to the construction of the regime for climate change mitigation and adaptation, an academic and policy discourse emerged in Western Europe and the United States on environment and security. Extensive studies of the complex linkages between environmental change and violence were undertaken (Homer-Dixon, 1999; Barnett, 2001). In very few, if any, instances could conflicts be directly attributed to environmental changes. The Darfur conflict in Sudan provides a recent much debated example of the sociopolitical consequences of desertification (Brown et al., 2007; Biello, 2009). It was an area in which defence establishments took a close and indeed self-interested view, as militaries attempted to justify their budgets in the era of the 'peace dividend'. In the late 1990s the North Atlantic Treaty Organization (NATO) commissioned a major study (Lietzmann and Vest, 1999) that sought to develop 'early warning' indicators for environmental changes that might drive conflicts. Evident reductions in Arctic ice cover and the ensuing struggle for territorial advantage and control of the new 'North West Passage' gave rise to the commonplace official assertion that climate change represented a 'threat multiplier' (European Council, 2008).

Such a view represented an extension of an entirely orthodox position on the meaning of security that continued to be defined in terms of the security of the state in the face of resource conflicts, forced migrations and insurrection. Alternative notions of security were also apparent in discussions of climate change, deriving from a shift in the 'referent object' of security policy. The object to be secured was transferred to individuals and social groups. More radical still were attempts to 'securitise' the climate itself (Adger, 2010) and to suggest that climate change represented a greater threat than terrorism.[8]

In April 2007 an official attempt was made to 'frame' climate change as a security issue. The, then, UK Foreign Secretary, Margaret Beckett, introduced a debate on the topic at the UN Security Council. It is probable that the intent was to provide some additional impetus to the Blair government's attempt, launched during the British joint presidency of the G8 and the EU in 2005, to lay the groundwork for a post-2012 agreement at the UNFCCC (Brown et al., 2007, p. 1144). The German government was similarly active within the G8 at Heiligendam in 2008, and in 2011 used its presidency of the Security Council to introduce another open climate debate. This was part of a wider diplomatic campaign to establish Germany's credentials as a prominent player in the resolution of environmentally related conflicts.[9] Neither of the Security Council debates had a substantive outcome, other than an agreement, in 2011, that the Secretary General would include climate issues in his reporting. The large number of national contributions was, nonetheless, revealing.

Detraz and Betsill (2010) have performed a content analysis of some of the official literature of the period, including the 2007 debate in the UN Security Council. They found that two central discourses emerged.

One, as in the emerging view of many defence establishments, linked environmental change to armed conflict. However, in the wider UN system an alternative discourse, labelled 'environmental security', held sway. This located climate change within a broader sustainable development agenda and relied on a United Nations Development Programme (UNDP) definition of 'human security'. The two discourses also reflected an institutional competition between the Security Council and the General Assembly, Economic and Social Council (ECOSOC) and the whole international environmental architecture that had been erected since the landmark UN Conference on the Human Environment held at Stockholm in 1972. The United States, Britain and France, along with other developed world members of the Security Council, were generally in favour of expanding the remit of the Council to include climate-related issues. In the 2011 debate, US representative Susan Rice was passionate in her demand for action. There was some precedent for including matters that would at first sight appear to lie outside a strict interpretation of the Council's mandate, relating to '... any threat to the peace, breach of the peace or act of aggression' under Article 39 of the Charter. Since 2000 the Council had debated and, in 2011 passed, Resolution 1983 on HIV/AIDS (human immunodeficiency virus/acquired immune deficiency syndrome). But the terms of the Resolution were quite strictly limited to the ways in which HIV/AIDS affected 'activities pertinent to the prevention and resolution of conflict'. The suggested terms of a climate resolution would have allowed the Secretary General to consider preventative action and identify future climate-related conflicts. From the Small Island States there was a more urgent call for enforcement and the president of Nauru, who addressed the Council in 2011, wondered pointedly whether the permanent members of the UN Security Council (P5) would fail to take action if it were their territories under threat of imminent inundation (UN Security Council, 2011).

Opposed to what was seen as 'mandate creep', detrimental to the proper responsibilities of the General Assembly and ECOSOC, were Russia and China, along with representatives of the Non-Aligned and G77. They claimed that the 'environmental security' discourse was appropriate rather than the alternative conflict perspective. Climate change was primarily a question of sustainable development rather than action to deal with threats to peace. Some speakers pointed out that the Security Council was dominated by big powers with its procedures far from transparent. There was suspicion that the engagement of the Council might be a way in which the developed countries proposing it could seek, once again, to evade their responsibilities to make emissions reductions under CBDR-RC and to avoid responsibility for development and adaptation funding. The July 2011 session ended with the agreement of a statement which simply allowed the Secretary General to refer to climatic drivers of conflicts by way of 'contextual information' in his reporting to the Council (UN Security Council, 2011).

Climate and population

In marked contrast to the UN security discussions, governments have been conspicuously silent on questions of population, consumption and climate change. From the time of Malthus, the consequences of an ever-rising human population have been a source of controversy. In the 1970s the issue was framed by commentators such as Ehrlich (1968) in terms of the limits to food production. Currently, public discussion has shifted to the manifold consequences of climate change, including loss of fresh water and the potential collapse of agricultural and maritime resources. In the most extreme version, a projected increase of global population to some ten billion by the middle of the twenty-first century entails the end of human society as we know it, a fate that appears all but inevitable (Emmott, 2013). Others point more optimistically to the possibility that a 'demographic transition' will occur to alleviate the problem (Pearce, 2011; Dorling, 2013).

In the simplest terms there may seem to be the most evident relationship between climate change and human population growth in the age of the 'Anthropocene'. This widely used concept encapsulates the view that human activities are now altering the planet, leaving markers in a way comparable to the natural changes that marked previous geological epochs including the Holocene (denoting the 11,000-year period since the last ice age). Involved are a cluster of changes such as ocean acidification and species loss directly associated with the changing chemical composition of the atmosphere.[10] The curve of rising GHG emissions over the last 200 years tracks that of population growth. In 1800 there were around one billion human beings, by 1900 1.65 billion and by 1960 three billion. In the mid-1960s human population was rising at an annual growth rate of 2.1 per cent, but the rate has since declined sharply. The demographic consequences were still substantial. By 2000 the global population stood at six billion and in 2013 at around seven billion.[11] Statistical analyses of historical data have confirmed that there is a proportional relationship, even though GDP increases may have a greater effect and the impact of population growth may be most significant in low-income societies (O'Neil, B.C., 2009, pp. 83–4). This leads to the general conclusion that:

> Slower population growth would lead to lower emissions, making the climate problem easier to solve. And slower growth would likely make societies more resilient to the impacts of climate change. (Ibid., 82)

The reference to resilience highlights the way in which high rates of population growth compound the problems of poor and vulnerable societies afflicted by the ravages of disease and food and water shortages associated with a changing climate. It is frequently the case that areas of high population growth and extensive vulnerability to climate change overlap.

On closer inspection the relationship between GHG emissions and demographic change is quite complex, as reflected in the scenarios built by IPCC and other researchers. Economic growth rates, average incomes per capita, the changing demographic structure of populations and varying assumptions about the carbon intensity of energy production are all significant variables. In the IPCC emissions scenarios population growth is positively associated with high emissions, along with increases in gross domestic product (GDP) (Nakićenović et al., 2000). More recent modelling relates income levels and carbon intensity to different projected levels of population growth. The findings do not point directly to the overriding significance of population changes but '... highlight the combined importance of both slowing population growth and reducing per capita CO_2 emissions to stabilise the global climate' (Royal Society, 2012, p. 81).

It is sometimes claimed that the population dimension of climate change is wilfully ignored. This cannot be the case in public discussion, but it is true that there is no serious framing in intergovernmental politics of climate change driven by population increases. On occasions it is mentioned in official speeches, but there has been no attempt to organise international cooperation around the stabilisation of population growth in relation to climate change. This may be because of inherent problems with calculating the figures and establishing a consensus on exactly what an international population regime could accomplish. Much more likely is that the issue is ignored because of its politically toxic character. It can give rise to highly divisive discussion of what Garrett Hardin (1974) once described as 'lifeboat ethics' – the view that the provision of aid to the hungry people of the developing world simply compounded the problem by creating a population-driven tragedy of the commons. Any attempt by developed world politicians to call for population reduction will tend to founder on objections related to the huge gap between the climate impact of rich and poor people.

Actual attempts at population control, however laudable the motives, raise profound human rights objections, whether in Indian sterilisation programmes or the Chinese 'one child' policy. International programmes on reproductive health and contraception have also proved capable of uniting what would, in other circumstances, be the most unlikely of oppositional coalitions, including the Catholic Church, Islamic authorities, the religious right in the United States and radical feminists. For all these reasons population issues are extraordinarily difficult for the UN system to handle and have been notably absent from, for example, the Millennium Development Goals. The 1994 UN Conference on Population and Development did consider the issue, primarily in terms of general human welfare but also environmental sustainability. Some progress was made in terms of a non-binding agreement on reproductive health, but very little has subsequently been achieved. The likelihood that an international population policy-related framing of the climate problem will emerge is, therefore, close to zero, whatever the evidence brought forward by the IPCC or other bodies.

Globalisation and consumption

Clearly it is not so much population *per se* but associated patterns of human consumption that are critical. Since the birth of the UNFCCC the economic processes of globalisation have accelerated, entailing, in particular, the displacement of production processes. Shipping containerisation and the development of air freight have been integral to globalisation, giving rise to increases in emissions alongside the facilitation of rising patterns of, largely unsustainable, consumption. An expanding global economy based on fossil fuel use and intensive agriculture must constitute the primary driver of emissions growth. Globalisation poses a particular problem for the territorially based allocation of responsibility for GHG mitigation under the UNFCCC. There is now substantial evidence of major emissions transfers from developed to developing countries as part of a globalisation process which leads to a 'spatial disconnect between the point of consumption and emissions in production' (Peters et al., 2011, p. 5). This casts very serious doubt on the validity of IPCC and UNFCCC accounting rules based on territorial emissions. Developed countries such as the United Kingdom (UK) have been able to report a stabilisation of emissions, while consumption has increased based on the falling price of manufactures associated with the fast-rising GHG emissions of China and other emergent economies (House of Commons Energy and Climate Committee, 2012). Measures to restrict emissions in developed countries through carbon-trading, which serves to increase energy prices, can compound the problem by promoting 'carbon leakage': the displacement of carbon-intensive industry offshore.

If the climate problem were to have been addressed from the standpoint of ecological holism, or even economic efficiency, the current territorially based arrangements and fragmented global architecture would hardly provide a rational solution. A more effective design would need, instead, to address the forces that drive the upward trend in emissions and the destruction of sinks. These are, potentially, within the control of governments, but such control has been willingly relinquished through the enabling of globalisation and the pursuit of a particular form of economic growth as the primary goal, even the test of legitimacy, for most governments. A critical enabler of economic globalisation has been the General Agreement on Tariffs and Trade/World Trade Organization (GATT/WTO) system, dedicated to trade expansion and economic growth. Economic orthodoxy and the prevailing view of trade specialists at the WTO is that there is no necessary relationship between trade expansion and environmental degradation, as long as the environmental externalities of production are factored into prices charged. The typical framing of the climate problem is that free trade and the protection of the physical environment go hand in hand. 'In the (1995) Marrakesh Agreement establishing the WTO, members established a clear link between sustainable development and disciplined trade liberalization' (WTO, 2013). Both the WTO and UNFCCC subscribe to the norms of a deeper international order that promotes the virtues of trade liberalisation and marketisation. Article 3(5) of the 1992 Climate Convention is clear on this when it speaks of cooperation to:

> ... promote a supportive and open international economic system that would lead to sustainable economic growth and development in all Parties, particularly developing Parties, thus enabling them to address the problem of climate change.

Furthermore, with particular reference to trade:

> Measures taken to combat climate change, including unilateral ones, should not constitute a means of arbitrary or unjustifiable discrimination or a disguised restriction on international trade.

WTO approaches to climate change stress the benefits of action to remove barriers to trade in environmental goods and services, citing, for example, a range of mitigation and adaptation technologies, identified by the IPCC, 'that can assist in the challenge of climate change' (WTO, 2013). Similarly, there are indirect benefits to climate change mitigation and adaptation from negotiations on agriculture and market access for non-agricultural goods. The elimination of agricultural subsidies will lead to the more efficient allocation of resources and increased trade opportunities and income for developing countries, which can reduce their vulnerability to climate change. The predictability of WTO commitments may offset climate uncertainties and 'ensure that developing countries do not suffer disproportionately from the negative impacts of climate change' (ibid.).

There is no conception, here, that the increasing scale of consumption and energy use associated with the progress of globalization may bear some relationship to rising GHG emissions. On the contrary, the only evident contradiction between the activities of the trade and climate regimes lies in the way in which the provisions of the latter may be at odds with the non-discriminatory rules of the former. Ensuring the 'harmonious co-existence' of trade and environmental rules has, thus, been a preoccupation of the WTO's Committee on Trade and Environment since its creation.

One reaction has been to accept the inevitability of population and economic growth with consequent burning of hydrocarbons and destruction of sinks. It is thus that the necessity of adaptation and the possibilities of compensation for climate loss and damage have come to occupy a much more significant place, alongside mitigation, in UNFCCC discussions considered in detail in Chapter 3. A much more radical alternative is to re-frame the problem in grand technological terms.

Geoengineering

If the age of the 'Anthropocene' involves human alteration of the physical characteristics of the planet, then, it may be argued, human intervention can also rectify the situation. The ultimate expression of what John Dryzek (1997) has termed the 'promethean' approach to environmental problems is geoengineering, '… a deliberate large-scale intervention in the earth's climate system in order to moderate global warming' (Royal Society, 2009, p. 1). Various possibilities have been mooted that fall into two broad categories, carbon dioxide removal (CDR) and solar radiation management (SRM). Carbon dioxide removal could involve a range of techniques including re-afforestation, various forms of land-use management and carbon capture and sequestration. Alternatively, there is the possibility of ocean fertilisation, requiring modification of the photosynthetic layer of micro nutrients to increase their uptake of carbon and ultimately deposit them on the seabed. Solar radiation management is based on different principles. Attempts could be made to change the surface albedo (a measure of reflectivity) of the planet leading to a direct cooling effect. Methods could involve increasing the brightness of the roofs of human settlements, positioning mirrors in desert areas, growing more reflective crops or increasing cloud cover (Royal Society, 2009, pp. 24–8). Another more certain, but much riskier, approach would be to inject the stratosphere with aerosols in imitation of the observed cooling effects of volcanic eruptions. Probably the most exotic of geoengineering proposals would rely on space-based systems to reduce the amount of incoming solar energy. Suggestions have included placing reflectors or sunshades in a variety of orbital positions. The costs and uncertainties, not least with regard to their impact on the earth's climate systems, remain prohibitively high – such as to render them an unrealistic proposition in the immediate future (Royal Society, 2009, p. 33). Also as Humphreys (2011, pp. 105–8) demonstrates, they raise novel and potentially difficult international jurisdictional problems in terms of the utilisation of the global space common.

There were nineteenth- and twentieth-century discussions of geoengineering in terms of weather modification or even the potential for weaponisation, leading to the 1977 ENMOD Treaty (Convention on the Prohibition of Military or Any Other Hostile Use of Environmental Modification Techniques). This outlaws the use of 'military or any other hostile use of environmental modification techniques having widespread, long-lasting or severe effects as the means of destruction, damage or injury to any other State Party' (Article 1.1). However, this should not '... hinder the use of environmental modification techniques for peaceful purposes' (Article III.1). Such a use had already been considered in a 1965 US President's Science Council report that provides the first apparent instance of such a technology-centred framing of the climate problem, in which 'the possibility of deliberately bringing about countervailing climatic changes needed to be thoroughly explored' (cited in Hone, 2013). This suggestion was taken up in a number of studies during the 1970s, 'However, in the 1980s and 1990s the emphasis of climate change policy discussions shifted to mitigation, primarily due to the efforts at the UN level to build a global consensus on the need for emissions controls' (Royal Society, 2009, p. 4). There was also concern, not just about the risks and uncertainties inherent in this under-researched area of human endeavour, but also the 'moral hazard' posed by any serious consideration of techniques that could encourage policymakers to avoid making what appeared to be politically costly commitments to GHG reduction.

As the prospects for a 'post-2012' climate agreement to succeed the Kyoto Protocol appeared to diminish, geoengineering made a renewed appearance on the public stage with the UK Royal Society Report (Royal Society, 2009) and a joint enquiry by the Committees of the House of Commons and the US House of Representatives. The former called for the examination of regulatory arrangements and expressed the need to 'push geo-engineering up the international agenda' (Humphreys, 2011, p. 114). Some geoengineering options, involving land-use changes, re-afforestation or carbon capture and storage are compatible with the existing climate regime complex and could be integrated into offset and trading arrangements at some future date. Others are not. The use of stratospheric sulphate aerosols would pose a serious threat to the Montreal Protocol regime for the protection of the ozone layer and would hardly be compatible with its role in GHG reductions. Space-based SRM is associated with a high level of potential risk including, possibly unintended, damage to the climate system. However, they differ from more mundane CDR in their hypothetical promise of effective and relatively quick countervailing action to reverse climate change (Royal Society, 2009, pp. 58–9). For such reasons commentators tend to regard them as the last resort that needs to be examined and held in reserve if the current regime fails when the 'option of geoengineering could look less ugly for some countries than unchecked changes in the climate' (Victor et al., 2009, p. 76).

Conclusions

The most obvious conclusion that emerges from a review of the 20-year history of the climate regime is the extent of institutional (perhaps more accurately organisational) inertia, coupled with the continuing stability of initial framings of the climate problem. Inertia contributes to fragmentation because of the political difficulty of closing down or re-organising parts of the UN system. There are many financial and politically self-interested incentives for governments to maintain and adapt existing organisations rather than to innovate. This leads to that institutional path dependence that is a consistent theme in the international cooperation literature (Keohane, 1984; Aggarwal, 1998; Keohane and Victor, 2010). General examples of such inertia are provided by the persistence of a Security Council membership reflecting the great power constellation of 1945; and by the continuance of the G8 and G20. The G8, which has taken up climate and energy issues, has a membership representative of the industrialised nations of 1976, with the subsequent addition of Russia in 1997. It was widely expected, around 2008, that the G20, a much more representative group of economically significant countries, would supersede the G8. It has yet to do so, moreover it should be remembered that G20 membership is, itself, frozen as at the date of its creation in 1999. One implication is that the organisational architecture seems to have become increasingly out of kilter, not only with the framings of climate change in the scientific and activist communities, but also with a rapidly shifting economic and geopolitical environment.

The institutional stasis of the UNFCCC emerges most clearly when considered alongside the geopolitical changes occurring parallel to its genesis and development; and the remarkable emergence of a globalizing economy over the same years. This is indicative of the obsolescence of a system based on national territorial emissions; and the failure to integrate rising international maritime and aviation emissions. Above all, the regime has had to grapple with its division into developed and developing countries under the common but differentiated responsibilities principle which, to put it mildly, no longer represents the 'respective capabilities' of the Parties. There are, of course, both principled and highly self-interested arguments for continuing the division between Annex I and the rest, with the effect that, despite some recent movement, a realignment of responsibilities from those set down in 1992 still appears, as will be demonstrated in Chapter 3, a daunting task.

In part, this state of affairs can be explained in terms of organisational self-interest, with supporters, secretariats and even delegates and client groups having a concern to maintain the status quo. Preservation of the existing organizational architecture also maps on to the intragovernmental division of labour. A concern with the niceties of organisational boundaries was clearly evident in the UNFCCC's relationship to the Montreal Protocol, whose secretariat was provided by UNEP, an option opposed by a G77 General Assembly majority when setting up the UNFCCC. That boundary is now being blurred, given the recent US–China agreement to use the Montreal apparatus for HFC–GHG reduction.

The relationship between the UNFCCC and the ICAO and IMO may seem to exemplify respect for the mandates of existing organisations. Yet the furore over the EU's attempt to treat aviation emissions differently, after decades of inaction by the ICAO, also illustrates the intensity of the political and economic interests engaged when attempts are made to shift the established order. It is also evident that the climate regime architecture reflects the political choices of major states as they compete for advantage or 'forum shop' for those organisations most congenial to their interests. G77/China had an evident interest in sticking with the UNFCCC because of its central principle of CBDR-RC and its links to the General Assembly and ECOSOC, where developing world voting strength can predominate in a universal organisation. This emerges clearly from the debates arising from moves in 2007 and 2011 to place climate change on the agenda of the Security Council. The United States sponsored what it saw as its own alternative to the much derided Kyoto Protocol (the Asia Pacific Partnership on Clean Development and Climate) at an APEC meeting in 2005. Since then there has been a string of other initiatives creating fora at various removes from the UNFCCC, such as the Major Economies Meeting and the Major Economies Forum. These serve, among other things, to emphasise that the United States is not simply an obstructionist power in climate change politics. Another example is the 2012 Climate and Clean Air Coalition to Reduce Short-lived Pollutants – including black carbon, methane and HFCs. The coalition, which includes the United States, Canada, Mexico and Bangladesh, operates within the framework of UNEP. Crucially it describes itself, in implicit contra-distinction to the UNFCCC/Kyoto approach, as 'cooperative and voluntary in character' (UNEP, 2012). Complexity is extended by the habit of other organisations to 'bandwagon' by including the 'master' issue of climate change in their remit (Jinnah, 2011). The all-embracing character of the problem lent this exercise some plausibility, even if it did not result in a great deal of positive remedial action.

The underlying scientific framing of the climate problem has remained remarkably consistent, with growing certainty as to the anthropogenic causes of the enhanced greenhouse effect. Serious socioeconomic analysis has displayed a similar consistency. In 2000 the IPCC's scenario builders summarised the main drivers in ways consistent with both previous and subsequent work, reinforcing '... our understanding that the main driving forces of future GHG trajectories will continue to be demographic change, social and economic development and the rate of and direction of technological change' (Nakićenović et al., 2000, p. 5). These 'natural frameworks', to use Goffmann's terminology, provide us with the closest approximation to the 'brute facts' of climatic change. However, they are hardly addressed by the international climate regime, despite the linkages with the scientific community through IPCC reporting and the Subsidiary Body for Scientific and Technological Advice (SBSTA). There are political and policy-related framings that do reflect the underlying science – discourses on sustainable development and on technical innovation, perhaps. However, the dominant political frames tend to underpin the existing fragmented institutional architecture. Attempts to re-frame the climate issue as a high priority security concern at the UN founder on suspicions by the majority of governments that this is merely a way for the developed world to evade its historic climate responsibilities. Concerted international action on demographic change has been ruled out, even if it might be desirable, by the rise of a peculiarly religious perspective in many otherwise diverse countries. This could not have been predicted around the time of the birth of the climate regime. Most significant of all has been a framing of economic and political life that has gained near universal acceptance since the ending of the Cold War; and which amounts to a central principle of the prevailing international order. This, of course, concerns the necessity of open market capitalism, the inadmissibility of government interference and the central importance of rising levels of economic growth and individual consumption. An essential contradiction is evident, both in policy and organisational terms, between the institutions devoted to economic growth and climate change mitigation.

These examples raise the perennial question of how political and economic interests in the creation or maintenance of organisations and initiatives may be distinguished from framing ideas. Marxist scholars would not admit this distinction. For them framings are essentially part of an ideological superstructure that cannot be detached from its material base. Hence, patterns of accumulation dominate the ways in which climate change has been framed. Issues of consumption and challenges to the existing neo-liberal political and institutional order would gain little traction in international climate politics. This interpretation could be supported by reference to the continuing insistence on market-based solutions in the climate regime complex and even in the enthusiasm for technological interventions that do not involve fundamental re-ordering of economic priorities. On the other hand, there are many aspects of the system that cannot really be explained in this way – including the uncomfortable consensus of the international scientific community on the anthropogenic bases of climate change and the many other framings of the climate problem. Constructions can exist independently of interests and have played a significant part in setting up a fragmented institutional architecture which is surprisingly resistant to change. It is within this setting that national and organisational interests are pursued and defended. Max Weber, writing in 1913, described the position as well as anyone:

> Not ideas, but material and ideal interests, directly govern men's conduct. Yet very frequently the 'world images' that have been created by 'ideas' have, like switchmen, determined the tracks along which action has been pushed by the dynamic of interests. (Weber, 1948, p. 280)

3
The UNFCCC Regime

The fragmented nature of interstate regulatory activity on climate change inevitably casts some doubt upon the continued significance of the UN's Climate Convention. It aspires to play a central coordinating role but is confronted by a growing array of sometimes unrelated, and usually unregulated, transnational and private governance activities (IPCC, 2014a). In the light of these circumstances, the devotion of an entire chapter to the intricacies of the UNFCCC requires some justification. Analysts have disagreed on the centrality of the Convention. For Keohane and Victor (2010) it remains at the core of the climate regime complex, but for Abbott (2012) it is one among many relevant intergovernmental, transnational and civil society entities. Where the UNFCCC sits in relation to present and future climate governance is a vitally important and unresolved question, but is not one posed in this book. Instead the focus is upon international climate politics, where attention remains fixed upon the Convention. This is despite those attempts, discussed in the previous chapter, to avoid, or even subvert, the UNFCCC. Most of these have been orchestrated by developed world governments. But the overwhelming majority of state Parties value the UN climate regime, because it is open to their influence and because they have development needs that may potentially be met within its expanding activities. In this sense, the regime is part of an underlying North–South bargain expressed in the Rio Earth Summit's concept of sustainable development. Nowhere is this more evident than in growing awareness of the necessity of properly funded adaptation to climate change impacts. This is something that has virtually no place among the many innovative mitigation activities beyond the UNFCCC regime.

The UNFCCC provides the legal framework for a commons regime. That is to say it represents an attempt by the international community to govern spaces beyond direct sovereign jurisdiction. In this sense the global atmosphere is one of four global commons, the remaining three being the oceans and deep seabed, Antarctica and outer space. Commons regimes differ from other attempts at global and trans-boundary environmental governance because they are designed to avoid what Garret Hardin (1968) famously described as 'tragedies'. Commons tragedies arise because there are short-run individual incentives to over-exploit a shared, but unregulated, resource which, unless checked, ultimately leads to collective ruin. In the case of climate change, the emission of excessive amounts of greenhouse gases and the destruction of sinks, while allowing short-term profit, leads to the loss of climatic stability with all its associated dangers. The integrity of the planetary atmosphere and climate has been described as the ultimate public good – that is something that cannot be provided through the operation of markets alone. The Stern Review (2007) characterised the climate problem as the world's greatest market failure. The point is that climatic stability has to be secured by the action of public authority. No such central authority exists in a decentralised system of sovereign states and therein lies the essential problem for international cooperation – the provision of 'governance' in the absence of government. It is a conclusion of Hardin's analysis that the avoidance of commons tragedies is impossible without the division of a shared common resource into 'enclosed' private property. For the global atmosphere this is not only a physical impossibility but there is no world government to enforce property rights and responsible behaviour.

Against what amounts to a counsel of despair, is the alternative view of commons governance championed by the work of Elinor Ostrom (1990) and her collaborators. Emerging from intensive study of large numbers of local commons institutions is the finding, contrary to Hardin's assumptions, that individual actors can build institutions and voluntarily regulate what remains a common resource. By such means have many local commons tragedies been avoided. It is a huge and uncertain step to transfer findings that apply to small face-to-face communities to a global scale, but there are several intriguing similarities. They at least provide some guidance to the institutional requirements of successful commons governance. There will need to be shared understandings of organising principles and the consequences of failure, along with means whereby neighbours can monitor and sanction each other's behaviour.

The institutional equivalents of local commons governance, at the international level, have been analysed, in the IR literature, as regimes. The regime concept first came to prominence in the aftermath of the global monetary crisis of the early 1970s in response to the question of what would replace the Bretton Woods monetary arrangements, based upon fixed dollar parities, that had underpinned the post-1945 growth of the western economies. The concept of a regime as a means of understanding and comparing the institutions of international cooperation was taken up by the dominant liberal institutionalist school of research and writing on international environmental problems. There are various other possible ways of describing international institutions, and regime categories overlap and are often inadequate. However, to avoid 're-inventing the wheel' and to facilitate comparison, they are used here to assist an analytic description of the UNFCCC and its evolution. In the classic statement provided by Krasner (1983, p. 2) and his colleagues, regimes comprise:

> ... sets of implicit or explicit principles, norms, rules and decision-making procedures around which expectations converge in a given area of international relations. Principles are beliefs of fact, causation and rectitude. Norms are standards of behaviour defined in terms of rights and obligations. Rules are specific prescriptions or proscriptions for action. Decision-making procedures are prevailing practices for making and implementing collective choice.

These are frequently difficult to disentangle and some analysts simply refer to norms of behaviour. There was discussion in the Intergovernmental Negotiating Committee (INC) as to whether 'principles' should figure at all in the UNFCCC text, with the United States resisting on the grounds that they might infer a legal obligation. Modifications were introduced to meet this concern by including a *chapeau* to Article 3 stating that Parties would 'be guided inter alia' by the principles (Bodansky, 1993, pp. 501–2).

There are certainly foundational beliefs of fact that underpin the regime plus central distributive principles and normative injunctions that determine who is to be responsible for taking action. The question of the differentiation of responsibilities and equity has been in contention throughout the life of the regime. Equally problematic have been the design principles of the regime in terms of 'top down' targets and timetables as opposed to less onerous 'bottom up' approaches. There are also important understandings, not always codified in treaties, as to financial obligations between North and South and the balance between mitigation and adaptation. Principles and norms are significant because regimes are said to change when these shift. The extent to which the regime has managed to change over two decades will be considered below and, to assist the reader, a chronological overview of the regime's evolution is provided in Table 3.1. The climate regime has also amassed a major corpus of rules. Those involving information, monitoring, review and means of enforcement are of great importance to the success of a commons regime because they will determine the extent to which neighbours will trust each other and be assured that other users

Table 3.1 Chronology of the UNFCCC regime

1992	Convention open for signature at Rio Earth Summit
1994	Entry into force
1995	CoP I Agrees the **Berlin Mandate** for a Protocol
1996–7	AGBM meetings draft a Protocol
1997	CoP 3 **Kyoto Protocol** agreed. Differentiated commitments for Annex I Parties totalling a 5.2% emissions cut for 6 greenhouse gases by 2008–12. Flexibility mechanisms: emissions trading, JI and CDM
2000	CoP 6 Hague EU–US disagreement
2001	US denounces Kyoto signature CoP 6 bis Bonn developed detail of Kyoto Protocol CoP 7 Marrakesh, agreed final terms of Kyoto Protocol
2005	Kyoto Protocol enters into force, EU Emissions Trading System commenced CoP 11/CMP 1 Montreal starts work on second phase of Kyoto Protocol AWG-KP
2007	CoP 13/CMP 3 Agrees **Bali Plan of Action** and sets up AWG–LCA convention track
2009	CoP 15/CMP 5 Copenhagen – **Copenhagen Accords**
2010	National pledges submitted to Secretariat CoP 16/CMP 6 Cancun formalises Copenhagen Accords, launches Green Climate Fund and Adaptation Framework
2011	CoP 17/CMP 7 Durban – **Durban Platform** – launches WG–ADP for a new agreement and agrees 2nd commitment period for Kyoto
2012	First Kyoto Commitment Period ends, Second begins, CoP 18/CMP 8 Doha
2013	CoP 19/CMP 9 Warsaw, discusses 2015 agreement and institutes 'loss and damage'
2015	CoP 21/CMP 11 New climate agreement under the Convention to be concluded

cannot 'free ride' on collective undertakings. The prevailing practices for making and implementing collective choice naturally involve not only the annual Conference of the Parties (COP) but also the subsidiary bodies and *ad hoc* negotiating groups that have been set up at various times to determine the regime's future path.

The UNFCCC has, since its inception, been based on a principle, whether seen as belief or a matter of scientific fact, that there is a need to achieve the '... stabilization of greenhouse gas concentrations in the atmosphere at a level that would prevent dangerous anthropogenic interference with the climate system ...' Art. 2). Through the influence of successive IPCC assessments, national scientific reports and campaigning by NGOs, the regime has been in a continuous dialogue with scientific findings on the extent, mechanisms and projected impacts of climate change. The design of a framework convention, building on experience with the 1979 Long-range Transboundary Air Pollution (LRTAP) and 1985 Vienna Convention, was to establish an institution which was open and responsive to changing scientific advice. In the LRTAP example there has been an iterative process leading to a succession of protocols dealing with different air pollutants. The Vienna Convention's Montreal Protocol (1987) has proved to be adjustable in regulating successive classes of stratospheric ozone-depleting substances. For the global climate, an unprecedented international scientific effort, centred upon the IPCC, has produced growing confidence as to the anthropogenic causes of ever-rising atmospheric concentrations of CO_2, although areas of uncertainty remain. These include, for example, the role of the oceans in the uptake of greenhouse gases and the precise location and magnitude of climatic impacts. The Convention set up a Subsidiary Body for Scientific and Technical Advice (SBSTA) to provide, as its name suggests, a continuous interface between climate and policymaking. A periodic review, linked to the publication of IPCC assessment reports, of what is termed 'the adequacy of commitments' has also been instituted (Decision 1/CP.16). The review is specifically tasked with consideration of the need to strengthen the long-term goal of the Convention in the light of evolving scientific evidence.

The anticipated proportionate response to increasing scientific understanding of the severity of the climate crisis has not yet occurred. An apparent unwillingness or inability of UNFCCC to take the scientific evidence seriously has been a source of continuous frustration, even rage, among environmental activists and those governments directly threatened by the impacts of increasingly severe weather events and rising sea levels. Progress in establishing a formal recognition of what would constitute 'dangerous anthropogenic interference with the climate system' has been terribly slow. In 1996 the EU pronounced that a mean temperature rise of 2 °C above pre-industrial levels represented the threshold of 'dangerous' change. The 2 °C threshold is usually associated with IPCC reports, although the latter body has 'never thus far attached a specific temperature threshold' to the concept 'dangerous anthropogenic interference' with the climate (UNEP, 2013, p. 2). Although widely accepted and discussed since then, it was only in 2009 that this figure was recognised in the Copenhagen Accord and subsequently formally agreed at the 2010 Cancun COP. For the Alliance of Small Island States (AOSIS), and many others, the 2 °C figure is unacceptable and the imperative is to allow mean temperatures to rise by no more than 1.5 °C.

The principles of 'equity' and 'common but differentiated responsibilities and respective capabilities' (CBDR-RC) have come to occupy a place at the heart of the regime. The exact meaning of the equity principle for the regime is difficult to determine. In the view of the Indian government it is 'an absolute and inalienable right that cannot be equated with, and is far beyond fairness' (Earth Negotiations Bulletin [ENB], 2013, p. 27). Its interpretation is potentially significant. 'Equity' has increasing profile 'as the distribution and pace of mitigation responsibilities increasingly mirrors a debate on access to ecological space' (ENB, 2011, p. 30) and its equitable use. It could also serve as a key distributional principle that referenced individual *per capita* as opposed to national emissions. These issues are at the core of arguments over global climate justice discussed in Chapter 5.

The CBDR-RC principle is closely related, but has found concrete expression in the categorisation of Convention Parties. The Parties to the Convention were divided into Annex I developed countries, charged with initial responsibility for taking the lead in emissions reductions and provision of development finance under Article 4.2, and the rest. In the INC negotiations no criteria for establishing the difference between developed and developing countries were established. The developed countries were simply listed. They comprised Organisation for Economic Cooperation and Development (OECD) members (identified in a separate Annex II) and the old Soviet Eastern bloc, defined as Economies in Transition and exempted from providing finance under Article 4.3. The composition of Annex I has come to seem increasingly outmoded as economic giants such as South Korea remain outside its ranks, but it has proved nearly impossible to add to its membership.[1] The Convention text was finalised in compromises agreed by the INC immediately before the Rio conference (Brenton, 1994, pp. 191–2). CBDR-RC wording does not appear to have loomed as large in the negotiators' minds as the related questions of whether to include emissions targets for developed countries and the arrangements for development funding. Both North and South supported the principle, but it was read in different ways. Developing countries stressed that 'common but differentiated responsibilities' reflected the culpability of the developed world, while the latter understood it as a commitment to take the lead because of their (then) superior economic and technical capabilities (Bodansky, 1993, p. 503). In this respect it is important to recall the adjoining phrase 'according to their respective capabilities' that is often omitted in representations of the principle. Subsequent contention over the extent to which it enshrines a permanent differentiated commitment has been one of the dominant themes in the evolution of the UNFCCC.

By 1994, when the Convention entered into force, the United States and its allies were already calling for developing country commitments but the compromise, on which the Berlin Mandate was built, ensured that these would form no part of the planned Kyoto Protocol.[2] It took two years to negotiate the terms of the Protocol, which mandated emissions reductions only by Annex I developed countries. Even before the signature of the Kyoto Protocol in late 1997, CBDR-RC and the absence of mitigation obligations for non-Annex I countries had become an issue in US domestic politics. The 'unfairness' of a system in which emergent economic rivals to the United States were required to make no emissions reductions became a central part of the Bush administration's justification of its rejection of the Kyoto Protocol in 2001.

The Kyoto Protocol finally entered into force through the efforts of the EU, and in the face of strong US opposition, in February 2005. As attention turned to what would replace the Protocol, on the expiry of its first commitment period (in 2012), the inevitability of mitigation action by developing non-Annex I countries, if there were ever to be an effective regime, was starkly apparent. In 2007 China had replaced the United States as the foremost (current) emitter of carbon dioxide and, since 2004, the International Energy Agency (IEA) has been predicting that by 2020 non-Annex I emissions would exceed those of the developed countries (Figure 3.1). In the same year the Conference of the Parties in its Bali Plan of Action (Decision 1/CP.13) recognised this by introducing the concept of Nationally Appropriate Mitigation Actions (NAMAs) for developing countries. This was part of a package deal

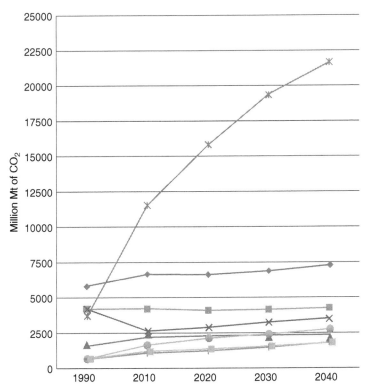

Figure 3.1 World energy-related CO_2 emissions
Source: EIA (US Energy Information Administration) 'International Energy Outlook 2013'
Available at: http://www.eia.gov/forecasts/ieo/table21.cfm. Accessed: 22/06/2014.

involving a 'shared vision' of comprehensive action in accordance with the CBDR-RC principle that acknowledged the importance of adaptation alongside:

> Nationally appropriate mitigation actions by developing country Parties in the context of sustainable development, supported and enabled by technology, financing and capacity-building, in a reportable and verifiable manner. (Decision 1/CP.13.1(b)(ii))

The importance of this new formulation should not be underestimated. Its use of words began to break down the rigid distinction between Annex I and the rest while holding out the possibility of differentiated commitments for developing countries according to their economic circumstances (ENB, 2007, p. 19). The intent of the Bali Action Plan was to pave the way for an agreement at the Copenhagen COP to be held in 2009. Shortly before the Conference both China and India set out their own mitigation actions in terms of decreases in the energy intensity of production, rather than quantified emissions reductions commitments. NAMAs were a key part of the Copenhagen Accord, which was the controversial outcome of the COP, and were later announced in a range of national mitigation pledges collected by the UNFCCC secretariat in the early part of 2010. The terms of the Copenhagen Accord were formally accepted in the following year at the Cancun COP, including developing country agreement on a range of diverse NAMAs (Decision 1/CP.16).

CBDR-RC and the Annexes have not been abandoned, but many Parties, including the United States, argue that the 'firewall' between developed and developing countries, that had been erected in 1992 and confirmed in the Kyoto Protocol, had been eroded. An important further step towards a new comprehensive climate agreement was taken in 2011 with the agreement of the Durban Platform (Decision 1/CP.17). This stated that negotiations should proceed towards an 'agreed outcome with legal force *applicable to all Parties*' (emphasis added). Arguments have continued over the nature of 'differentiation' between the Parties. The Umbrella Group of developed countries opposed mention of 'common but differentiated responsibilities' as the determinant of future obligations, and their G77 counterparts continued to insist upon it. At the Doha COP in 2012 the Umbrella Group and the EU expressed the view that 'Convention principles should be seen in an evolving context, noting the need to discuss further the principle of equity in terms of fairness and changing realities'. Developing countries stressed their opposition to 'any re-writing or re-negotiation of the Convention's principles' (ENB, 2012, p. 16) but the changes evident since Bali still seem to offer significant room for compromise on the construction of a new comprehensive regime. Nonetheless, the question of differentiating between the obligations and responsibilities of the Parties runs like a red thread through the elements of the Convention that will be discussed in this chapter, just as the negotiations within the Ad Hoc Working Group on the Durban Platform for Enhanced Action (ADP) have been 'permeated' by the question of how differentiation will be reflected in a 2015 Paris agreement (ENB, 2014, p. 43).

The decision on NAMAs, quoted above, references another key principle of the regime – that the 'Parties have a right to, and should promote sustainable development' (UNFCCC, Art. 3.4). This serves as an expression of an underlying understanding, central to environmental politics since the first UN environment conference in 1972 and reinforced by the Rio Earth summit, at which the UNFCCC was opened for signature, that there was a necessary relationship between environmental action and development. In the Convention text it appears as an obligation placed upon developed countries to provide 'new and additional financial resources, to meet the agreed full costs of developing countries in fulfilling their reporting obligations' (Art. 4.3). More broadly, the Convention recognises the need for development aid and technology. It is possible to regard the obligation of developed countries to provide the funds for assisting mitigation, adaptation and participation by the developing world as a regime principle beyond the precise wording in the text, as the reciprocal basis for any deal that may be struck on a comprehensive approach to greenhouse gas (GHG) mitigation. This has been implicit in the various offers of climate funding that have been made in advance of the Copenhagen COP and in preparation for a 2015 agreement.

Mitigation principles and rules

While CBDR-RC attempts to assign the relative burden of GHG mitigation, the nature of such commitments, and whether or not they should be internationally determined and enforced, has been a contentious issue throughout. It was the central subject of EU–US dispute during the INC negotiations, leading to the compromise embodied in Article 4.2(b) of the Convention, whereby the Parties would merely 'aim' to reduce their emissions to 1990 levels by 2000. The achievement of the Kyoto Protocol was to put in place an internationally agreed and binding set of mitigation commitments for six GHGs, to be achieved, in line with the CBDR-RC principle, by developed Parties by 2008–12. The national commitments were differentiated, leading to an overall 5.2 per cent reduction against a 1990 baseline. National commitments were operationalised as Quantified Emissions Limitation or Reduction Objectives (QELROs). Kyoto may, therefore, be described as a 'top down' agreement with a defined international reduction target and quantified and binding commitments by its developed (Annex I) Parties.

The Protocol, agreed in outline in 1997, was only developed into a detailed instrument, capable of ratification, in a process organised by the EU, in the face of US opposition, during 2001 (COP 7 Marrakesh Accords). After a demanding quest for the necessary number of ratifications, once again led by the EU, it entered into force in February 2005. Even before ratification, the ambitions of the Protocol were widely disparaged as inadequate to the task of reducing GHG emissions to the levels required if the assessment reports of the IPCC were to be taken seriously. Advocates of the Protocol could respond that, despite its limitations (exacerbated by the way in which developed Parties either failed to ratify or subsequently reneged on their commitments) it at least provided an international foundation upon which future progress in mitigation could be built. The Stern Review of 2007, and mainstream economic commentary, have stressed the importance of establishing a global carbon price to include the 'externalities' of fossil fuel use and to encourage alternative and climate-friendly investment. The Kyoto Protocol could thus be represented as an essential first international step in this direction.

Its architecture was both complex and path-breaking. At American insistence, an agreement on 'targets and timetables' was made contingent upon the acceptance of 'flexibility mechanisms' – Joint Implementation (JI), the Clean Development Mechanism (CDM) and emissions trading. These provisions were intended to provide economically efficient 'market-based' alternatives for governments which did not wish to simply impose restrictions or taxes on emissions in order to achieve their mitigation commitments. It is worth noting that the applicability of 'market-based' instruments is still a matter of dispute among the Parties and is by no means universally accepted.

National and international carbon markets have been created using the Kyoto rules – the principal international example being provided by the Emissions Trading Scheme of the EU. These are linked to the other offset mechanisms of the Protocol, CDM and JI. Both allow Parties to invest in emissions reductions projects in other countries and earn credits, Certified Emission Reduction Units (CERs) in the case of the CDM and Emission Reduction Units (ERUs) for JI. They can be traded or used to achieve the investing countries' own national targets. Each CER is equivalent to one tonne of CO_2 emissions avoided. The logic of the system is that it will encourage those developed Parties that are already energy efficient to achieve greater carbon reductions by investing money elsewhere – in the case of JI, in other developed countries, and in the case of the (very much larger) CDM, in carbon reduction projects in developing countries. Since 2006 the CDM has grown apace, with over 7,000 registered projects but also amidst accusations of fraud and sharp practice. Such a system has required unprecedented levels of institutionalisation and regulation, through a central transaction log and registry, along with a highly developed enforcement and compliance system to counter the evident opportunities for abuse. Alongside the flexibility mechanisms, Parties can also gain emissions credits by investing in land use, land use change and forestry initiatives (LULUCF). It should be recalled that the Convention covers both sources and sinks for GHGs, and the LULUCF sector involves both. LULUCF has proved controversial over the years because of the opportunities it might provide to avoid making actual emissions reductions. Hence, complicated accounting rules were devised that have been revisited in ongoing specialised negotiations.

The other sources of credits for Kyoto Parties are surplus 'assigned amount units' (AAUs). These occur when actual emissions are below annual targets and can be traded. Such dealings in 'hot air' have been a source of outrage among environmental activists, who point to the way in which Russia, for example, has accrued a large surplus of AAUs, its faltering economic performance having caused it to undershoot its predicted emissions.[3]

Although it was the progenitor of these 'market-based' systems of emissions mitigation, the United States never ratified the Protocol. It was the EU that reversed its previous reliance on 'command and control' regulation and embraced the new approach. By 2005 EU–US climate relations had descended to a new low as the former championed the Protocol while the latter denounced 'targets and timetables' and even questioned the scientific basis of the regime. The first Kyoto commitment period expired in 2012 and the terms of the Protocol (Art. 3.9) required that a successor should be the subject of international discussion by 2005. In the search for a post-2012 agreement it was readily apparent that the United States, and even some existing developed Parties, would not subscribe to a new 'top down' approach. On the other side, the G77 and China required that there be a second commitment period for Kyoto as a condition of their own participation in future mitigation actions. At the Bali COP in 2007 a procedural solution was found by splitting the negotiations along two tracks, the existing one on the future of Kyoto, in which the United States did not participate, and the other, a new working group on 'long-term co-operative action' under the Convention, in which it did.[4]

The two negotiating tracks were supposed to converge at the 2009 Copenhagen COP, at which a comprehensive post-2012 agreement was to be produced. In the event, the formal negotiations stalled and the COP produced something rather different in the Copenhagen Accord (CP/2009/L.7). A central feature of this agreement, between the United States and the new BASIC coalition of Brazil, China, India and South Africa, was its reliance on 'bottom up' pledges on emissions reductions, which had a purely national character. Annex I countries were to submit 'quantified economy wide emissions targets for 2020' while others would undertake NAMAs. Within the Accord the context of these undertakings was the provision of finance to developing countries and the creation of the Green Climate Fund. In early 2010 two lists of national offers were compiled as appendices to the Accord. The quantified pledges of Annex I countries differed widely in terms of their percentage reductions and associated baselines and there was even more variation among the NAMAs submitted by non-Annex I Parties. The Copenhagen Accord was formalised at the subsequent Cancun COP of 2010.

It is very unlikely that the various pledges submitted by 42 developed countries and 55 developing country parties will be sufficient by themselves to close the 'emissions gap' by 2020. The 'gap' is the difference between the emissions levels that will be achieved in 2020, if all commitments and pledges are achieved, and that which would be consistent with stabilising mean temperatures at 2 °C and 1.5 °C increase over pre-industrial levels. The question of its achievement will be considered in detail in Chapter 8.

The firmest action on mitigation has been taken by the EU and those other countries that have undertaken quantified reduction commitments (QELRCs) for a second commitment period, 2013–20, under the Kyoto Protocol.[5] This was a negotiating demand of the G77 but, as will be seen, various original Parties to the Protocol have refused to be involved in its continuation. Non-participating Annex I parties, including the United States and Japan, have submitted pledges of varying ambition which do not constitute binding commitments and are generally expressed in terms of reductions from historic baselines, usually 1990. For non-Annex I countries with development ambitions the situation is necessarily different. Their NAMAs are calculated in relation to 'business as usual' on their development trajectories – that is to say as a reduction against estimated future emissions levels. For China and India there are pledges to cut future emissions intensity – reducing the amount of carbon emitted per unit of GDP. Other developing countries have submitted a variety of nationally appropriate actions which may involve sectoral targets or even specific projects.

Post-2020 mitigation

It is relatively certain that the planned agreement for 2020 will not resemble the Kyoto Protocol. As we have seen, the absolute distinction between Annex I and the rest has been removed to the extent that mitigation actions will be an obligation for all Parties in an agreement 'applicable to all'. There is also an acceptance, even by the United States and the Umbrella Group, that the principle of CBDR-RC still pertains, but with 'national efforts ... differentiated across a broad range of parties' (EU, 2013, p. 2; United States Government, 2014, p. 1). The US chief negotiator in 2013 argued that the avoidance of 'top down' targets such as those in the Kyoto Protocol made it possible to maintain CBDR-RC through a flexible approach where countries could protect their development aspirations under a new agreement applicable to all (Stern, 2013, pp. 5–7).

In a critical departure from the principles upon which Kyoto was built, 'contributions' will be nationally determined. The phrase used in the ADP negotiations is 'Intended Nationally Determined Contributions' (INDCs). This means that the agreement will be 'bottom up' and constructed in terms of what Parties are willing to pledge, rather than 'top down' according to some agreed global target similar to that for developed countries contained in the 1997 Kyoto Protocol. An important compromise reached at the Warsaw COP, in late 2013, was that the language of 'commitment' would be replaced by that of 'contribution'. The EU, in line with its previous policy, had pressed for a comprehensive new protocol that would include legally binding national commitments. This proved unacceptable to India and other developing countries. The compromise wording which was eventually agreed left wide open the question of the precise legal nature of the obligations to be assumed under the 2020 agreement.[6] A previous compromise at Durban, in 2012, had contained a catch-all phraseology that the 2015 agreement should be in the form of a 'protocol, legal instrument or agreed outcome with legal force' (ENB, 2011, p. 28). This is not simply the result of developing countries wishing to avoid being forced into inequitable legal obligations that would compromise their economic prospects; it is also a concern for the United States, where the legal characteristics of 'contributions' are significant because of the difficulties in implementing commitments proposed by the Executive Branch that would require the approval of Congress.

The emerging agreement may bear some resemblance to the kind of 'pledge and review' mechanism proposed by Japan, but rejected during the preparatory INC negotiations on the Convention. Discussion in 1991 dealt with the legal nature of pledges, whether they should be unilateral or in response to a given international target. According to a Chatham House study group convened to consider the question, 'pledges should be *clear, significant and defined in such a way that undertakings can be verified*' (Royal Institute of International Affairs [RIIA], 1991, p. 5, emphasis added). Pledges would be the expression of national mitigation strategies, but concern was expressed about the expectation that developing countries would be required to produce pledges and incur costs 'to address a problem that they had played scarcely any part in creating'. For many, but not all, developing countries this remains the case. In other areas, uncertainty expressed about a future regime in 1991 has been replaced by a great deal of accumulated institutional experience – on offset mechanisms, accounting rules and the critical question of monitoring and verification.

Measurement reporting and verification (MRV)

Satisfactory rules for monitoring and verifying participant behaviour constitute a vital prerequisite of any effective commons regime. It will be necessary to establish the extent to which problems are being solved and targets met and, indeed, to identify the nature and extent of problems that imperil the commons. Without information, it will not be possible to understand the deficits in capability, which prevent full participation in the regime. Paramount will be the requirement to establish trust and to demonstrate the fulfilment of commitments, without which there will always be concerns over cheating. Such requirements were recognised and embodied in the 1992 Framework Convention. Indeed, the provision of information was the main obligation undertaken by all signatories.[7]

Over the lifetime of the Convention the development of MRV has been extensive. Annex I Parties submit regular biennial national communications on policies and measures and many other aspects of their response to climate change which since 2014 have been subject to an International Assessment and Review (IAR) process to promote comparability of reporting. They are also required to submit annual national inventories of GHGs which are subject to expert technical review. Developed country Parties to the Kyoto Protocol are also subject to additional reviews and there is a compliance mechanism which is justly described as 'among the most comprehensive and rigorous systems' to be found in any multilateral environmental agreement (UNFCCC, 2013, p. 32). It is necessitated by the need to prevent fraud and to maintain the integrity of carbon markets.

Special consideration has always been given to the needs of developing countries often lacking the capacity to fulfil the information-gathering requirements of the Convention. The reporting and analysis problems encountered can be very substantial. For example, in the case of Malaysia, a middle-income developing country, it took ten years to compile data on its situation in 2000. In the context of the 2020 agreement there is real concern that governments will be pressured into accepting NAMAs without a full understanding of their economic implications.[8] Ever since MRV was introduced for all parties in the Bali Action Plan (2007) there have been North–South arguments over sovereignty and the extent of funding for 'capacity building'. The sensitivity of the MRV issue is reflected in the agreed description of the International Consultation and Assessment (ICA) process for developing countries as 'non-intrusive, non-punitive and respectful of national sovereignty' (ENB, 2013, p. 6).

Review processes remain critical to the design of the post-2020 regime. Although Secretariat spokesmen have been at pains to avoid using the term 'pledge and review', the experience with the national pledges notified in 2010 reflects the difficulties likely to be encountered in assessing the adequacy of the diverse 'contributions' to a new agreement. There are calls from the EU and United States for a robust comparative element of international assessment which will allow parties to assess the sufficiency of global effort in aggregate and provide incentives for Parties to engage in strict implementation.[9] There is also the matter of when assessment should occur and the argument that this should be *ex ante*, before a 2020 deal is concluded. As always in such international agreements there is the lingering suspicion that rivals will take advantage of an agreement which is neither transparent nor subject to watertight verification.

Adaptation

Adaptation refers to the adjustment of ecological, social and economic systems to the actual or potential impacts of a changing climate. It involves the assessment of climate vulnerability and the means to plan, implement and fund necessary remedial action. All societies will face adaptation problems but the least developed will tend to be the most vulnerable, often lacking the means to preserve their economic and social fabric. This was recognised in the Convention (Art. 4.4) but downplayed in the sense that adaptation did not figure, alongside mitigation, as an Article 2 objective. This was reflected in initial funding arrangements and the Global Environment Facility (GEF) rules required global environmental benefits which precluded spending on adaptation (South Centre, 2011, p. 8). In 2001, it was agreed that the

Kyoto Protocol should have an adaptation fund which receives 2 per cent of CER returns. Adaptation achieved greater prominence as the effects of climate change became more evident, and was given equal weighting to mitigation in the 2007 Bali Action plan, formalised in 2010 in the Cancun Adaptation Framework. This involves the drawing up by developing countries, with an emphasis on least-developed countries, of National Adaptation Plans (NAPs). The intention is that these will be funded, initially, through the GEF and a dedicated adaptation fund. Developing countries have demanded that a prerequisite of their mitigation actions should be international funding of adaptation. In their view there is an enduring link between emissions reduction and development and adaptation finance. 'Only when finance was provided could a developing country be expected to carry out its pledge' (ibid., p. 11). They also argue that in a 2020 agreement the 'global challenge' of adaptation 'be addressed with the same urgency as, and in political and legal parity with mitigation' (UNFCCC, 2014, p. 2). Contrary to the wishes of developed countries, this would make adaptation and adaptation funding one of the 'intended nationally determined contributions'.

Provisions for 'loss and damage' are a recent addition to the adaptation framework, formalised in the Warsaw International Mechanism for Loss and Damage (2/CP19). The reference is to impact upon particularly vulnerable developing countries occasioned by 'extreme weather events and slow onset events' that cannot be prevented by any amount of mitigation. These arrangements, which are intrinsic to discussions of responsibility and justice in the regime, are further discussed in Chapter 5. For less-developed countries and Small Island Developing States (SIDS) it is important that 'loss and damage' provisions and funding are kept separate from other parts of the adaptation agenda, and that they should form part of a 2020 agreement.

Finance

That cooperation, reporting and action by non-Annex I developing countries are contingent upon the provision of financial aid, technology transfer and capacity building by their developed counterparts may be regarded as an operating principle of the regime. Article 4.3 of the Convention commits developed countries to provide 'new and additional financial resources to meet the agreed full costs incurred by developing Parties in fulfilling their reporting obligations and the incremental costs of their more general commitments'. Responsibility for providing the funds fell to the Annex II countries (Annex I minus the East European and Russian 'economies in transition').

A financial mechanism for resource and technology transfers was part of the Convention (Art. 11) but without any concrete arrangements. Controversially, the GEF of the World Bank was selected as its operating entity, although distrusted by developing countries as a body beyond their, or the COP's, control. The funding provided was limited and largely targeted at mitigation efforts.[10] The GEF's operations were also criticised for their lack of transparency and for the way in which World Bank indicators were deployed without consultation (Gomez-Echeverri and Müller, 2009).

Since the 1990s, as the scale of the overall task of responding to mitigation and adaptation challenges began to be apparent, climate change became a major part of the remit of development institutions and bilateral aid programmes. The sums required dwarfed those provided under the UNFCCC/GEF arrangements; they would only increase should both developed and large developing countries fail to take timely action. Thus, part of the Bali Action Plan of 2007 was the call for 'enhanced action on the provision of financial resources'. The response was agreement in the Copenhagen Accord, formalised at Cancun in 2010, for $30 billion 'fast start finance' donated between 2010 and 2012 and for the setting up of a longer-term dedicated Green Climate Fund under the Convention. The Fund was established with headquarters in South Korea and by the end of 2014 had been capitalised to the sum of $10.2 billion from developed countries. Its projected target was to raise $100 billion from public and private sources by 2020, but it remains unclear how this is to be achieved and whether such a sum will be adequate to the task. Some estimates predict that the sums needed by 2030 will be three times that figure (South Centre, 2011, p. 9). There are also major issues of transparency and 'additionality' surrounding climate funding in general and 'fast start finance' in particular. Whereas developed donors have apparently committed the promised 'fast start funds', there is uncertainty as to what percentage of them are actually new grant money as opposed to loans and repackaged aid. The extent to which developing countries can be confident of the future funding of climate action remains a major determinant of their participation in a comprehensive 2020 agreement.

Over the years a number of development-related mechanisms and programmes have been established within the UNFCCC, for example various technology transfer mechanisms and a Technology Executive Committee (TEC) charged with developing links with the funding agencies discussed above. Another long-discussed way in which developed countries could contribute to climate-related actions in the developing world is through Reduction of Emissions from Deforestation and Forest Degradation (REDD+). This approach to the inclusion of forest sinks and sources in the UNFCCC dates from a 2005 proposal introduced by Papua New Guinea, although the forestry issue had long been a staple of North–South discussions, involving the failure to conclude a forests agreement complementary to the 1992 Climate and Biodiversity Conventions. In fact, the forestry dimension of the UNFCCC is very underdeveloped in comparison to other international arrangements and to the large number of private and public forestry initiatives that have emerged elsewhere.[11]

After years of discussion a package, 'The Warsaw Framework for REDD+', was agreed in 2013 on institutional arrangements, principles, methodologies, monitoring and potential funding. It is to be stressed that this is still a framework rather than an operational system rewarding efforts to conserve the carbon in forests. Although the management of forests is a very significant part of both the climate problem and its solution it has proved to be an extremely difficult issue for the regime, resulting in the abandonment of attempts to negotiate a forest component of the Kyoto Protocol. The problems encountered in this sector are a subset of the broader difficulties of building a climate regime. How to establish that forestry actions are long term, not subject to misallocation and 'additional'? How, also, to ensure the environmental integrity of forests beyond simply ensuring that emissions are avoided and sinks preserved? These tasks might, after all, be accomplished by cutting down ancient woodlands and replacing them with fast-growing commercial plantations, with potentially dire consequences for biodiversity and local indigenous livelihoods.

REDD+ involves a North–South deal – 'In the context of the provision of adequate and predictable support to developing country Parties, Parties should collectively aim to halt and reverse forest cover and carbon loss' (Warsaw Framework for REDD+). The Global Climate Fund is supposed to fund REDD+ initiatives, but this makes them subject to the funding problems discussed above. Otherwise, there is the question of the extent to which markets should come into play and the potential for sharp practice when forestry offsets are created and traded – thus raising the issue of the validity of market mechanisms in an acute fashion. Additionally, there is the suspicion, among developing countries, that Annex I Parties may use support for REDD+ as offsets to avoid their emissions reduction obligations (BASIC, 2013a). Finally, REDD+ demonstrates the painfully slow process of rule development within the principles of the regime.

Decision-making procedures

The supreme body of the Convention is its Conference of the Parties (COP). It holds annual meetings, its slot in the international calendar being November–December. Since the entry into force of the Kyoto Protocol it has been conjoined with the Conference of the Parties serving as the Meeting of the Parties to the Kyoto Protocol (CMP). Thus, in a particular year, say 2015, there will be a Conference designated as COP 21/CMP11. COPs have become large and high-profile international events, attracting very substantial participation from global civil society. At COP 1 in Berlin in 1995 there were 2,900 participants, of which 757 were delegates. At COP 3 in Kyoto in 1997 the number of participants had risen to 6,000 (Yamin and Depledge, 2004, p. 31). The 2009 Copenhagen COP 15 represented something of a peak, with no less than 10,951 delegates and 13,482 other participants (Schroeder et al., 2012, p. 835). The 2013 COP 19 at Warsaw was on a more typical scale with 8,300 participants, including 4,022 government delegates (ENB, 2013, p. 1). Normally, more than half of registered participants are not accredited delegates and COPs have been enlivened, not only by a range of side events often with commercial, scientific or NGO sponsors, but also by sometimes flamboyant political protests in which NGOs, among their various other significant roles, serve as a kind of Greek chorus to the formal negotiations. In the last decade the availability of online video casts of Conferences, allied to social messaging, has expanded such opportunities. This may be viewed as a prominent example of the rise of a new, 'real time' interconnected engagement by global civil society, but it is difficult to gauge its impact on the course of negotiations.

The formal business of the COPs is conducted by governmental representatives who sit on the many committees and subgroups, including meetings of the subsidiary bodies, which convene for a fortnight below the level of plenary meetings of the Conference. While most of the business is conducted by officials, government technical specialists and representatives of special interests included in national delegations, the final few days are designated the 'High Level Segment', when ministers and even presidents and prime ministers put in an appearance. Plenary sessions of the COP can become very lengthy as many of the 196 Parties may wish to make formal statements of position alongside invited speeches from dignitaries such as the UN Secretary General and head of the IPCC. The presence of ministers and the need to conclude with a positive outcome frequently lead to last-minute negotiations, late-night sessions and the over-running of the conference deadline that have almost become a standard operating procedure of the COP.

The Presidency of the COP is held by a ministerial representative of the host country. At its very first meeting in Berlin in 1995 it was a position held by Angela Merkel (ENB, 1995, p. 9). The Presidency plays an important role in setting the agenda, in consultation with the national representatives elected on to the bureau of the COP, and in orchestrating negotiations when the Conference is in session. The occasionally high political visibility of the COPs should not obscure the fact that the attempt to negotiate new agreements and to conduct the business of the regime continues year round. Numerous bodies have been established under the Convention and Kyoto Protocol (see Figure 3.2) but the most important are the Subsidiary Body for Scientific and Technological Advice (SBSTA) and the Subsidiary Body for Implementation (SBI). The function of the SBSTA is evident from its title. It considers, at expert level, informational requirements and methodological issues. The SBI has a parallel role, specifically in the consideration of national communications that are mandated under the Convention. Both bodies meet regularly at mid-year in Bonn, the seat of the Convention Secretariat, as well as at the COP. Much attention will focus on the *ad hoc* temporary bodies set up to draft future agreements. The first was the AGBM (*Ad Hoc* Group on the Berlin Mandate) that negotiated the text of the Kyoto Protocol over eight meetings during 1996–7. The two *ad hoc* negotiating groups on the future of the Kyoto Protocol (AWG-KP) and the Convention (AWG-LCA) were mentioned earlier in discussion of the two-track approach adopted as part of the Bali Plan of Action. In the event the AWG-LCA's laborious negotiation of heavily square-bracketed text came to nothing as leading Parties agreed to the alternative Copenhagen Accord. Both groups were wound up at the Doha COP in 2012. Their replacement had already been launched in 2011, the ADP or *Ad Hoc* Working Group on the Durban Platform for Enhanced Action. It has met regularly to consider the terms of the new agreement for 2020 that will be presented at the 2015 Paris COP. These negotiations centre upon a laborious exercise in textual drafting. While there are plenary sessions of the ADP to take stock and to approve outcome documents, detailed work will be undertaken in contact groups which are 'open ended' in terms of participation and informal consultations which are not. Here the Conference Presidency, along with the Co-Chairs of the ADP, plays a significant role in organising meetings and determining the delegations that will be invited to participate.[12]

The Conference was supposed to agree to its rules of procedure, including majority voting on specified issues, at its first meeting (Convention Art. 7.3). However, there were objections from Saudi Arabia and other Parties fearing that they would not be able to cast

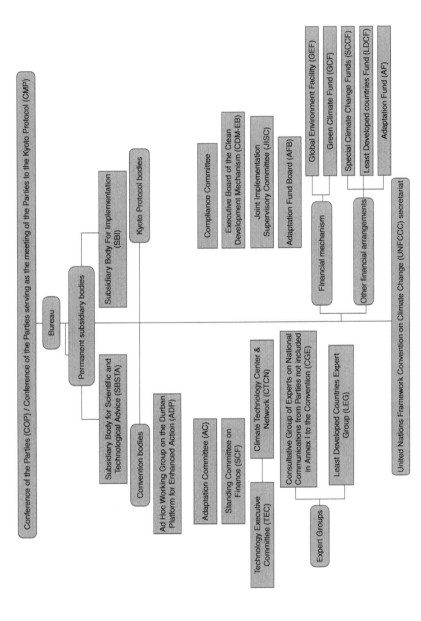

Figure 3.2 Organigram, UNFCCC regime. Acknowledgement. Reproduced from the UNFCCC Secretariat website, unfccc.int

vetoes on questions prejudicial to their interests. As a result, draft rules of procedure have been applied ever since – with the exception of rule 42 on voting (FCCC/CP/1962/2). It soon became clear, however, that the Parties were willing to act without a full consensus. They first proceeded on this basis in issuing the Geneva Ministerial Declaration of July 1996 (Bodansky, 2001, pp. 35–6). The 2009 Copenhagen Accord was an 'accord' rather than a formal act of the COP because there was no consensus on account of the objections raised by Venezuela, Cuba and other members of the Bolivarian Alliance (ALBA). More recently, objections by Russia, Belarus and Ukraine at the 2012 COP went 'unnoticed' by the chair, leading to retaliation at a subsequent SBI meeting.[13] The issue of whether Parties shall be allowed to block the will of a clear, even overwhelming, majority is likely to recur.

In common with other large multilateral gatherings, there is a continuing problem with the effective exclusion of many small and developing Parties from key informal discussions, and even from any meaningful participation, in substantial parts of the regime's work. There are important 'capacity' issues. Delegation sizes vary significantly, with most Parties only able to send a handful of delegates while the big players and host countries of the Conference are able to field delegations sometimes running to hundreds of personnel, with a range of expertise sufficient to cover the full span of the work of the COPs and subsidiary bodies (Schroeder et al., 2012) –

> ... many sessions take place in parallel, span a wide range of issue areas and continue into the night during the final 'push' for agreement at the end of a conference. As a result 'negotiation by exhaustion' constrains many smaller delegations much more severely than larger ones. (Ibid., p. 835)

One of the critical functions of interested environmental NGOs has been to attempt to close this 'capacity' gap, but developing Parties are still faced with the issue of whether it makes sense to deploy what are often a limited number of national experts at the international level, as opposed to the potentially more useful work that they could undertake at home.[14]

Compounding the capacity problem is the need to negotiate through informal meetings, 'drafting groups' and appointed 'Friends of the Chair'. They may be necessary in order to resolve difficult issues in private but their membership is necessarily selective. The most egregious example of exclusion occurred in the final cabal between the

United States and China, India, South Africa and Brazil at the 2009 Copenhagen COP that drafted the Accord. Since then there have been serious efforts to increase the transparency and inclusiveness of conference processes. Much depends on the willingness of the host Presidency of the COP to engineer informal processes that allow fuller and more balanced involvement. For example, the South African Presidency at Durban arranged a series of 'indaba' meetings that appear to have promoted agreement (ENB, 2011, p. 30). Nevertheless, distrust of the Convention's procedures remains, often expressed in demands for a 'Party driven process'.

Conclusions

The international climate regime has had a bad press over a long period. Climate 'gridlock' was predicted even before the signature of the Convention (Skolnikoff, 1990) and has been a recurring description (Victor, 2011). Skolnikoff (1990, p. 78) argued that the Convention that he expected to be negotiated by 1992 would most likely 'be an empty shell for many years' because of the high barriers to action and agreement and the public's unwillingness to commit to issues that were both 'costly and intangible'. As it turned out, relatively rapid progress was made after 1992, with the implementation of the Convention and the negotiation of the Kyoto Protocol. The price of initial agreement was to embed a North–South distinction at the heart of the principles of the regime, which was sustained in the formulation of the Protocol. This created many difficulties in implementing the Protocol and limited its potential effectiveness and acceptability, as rapidly changing economic conditions radically altered the 'respective capabilities' of the Parties. The rigid division of the world into Annex I and non-Annex I has proved particularly difficult to alter because it reflected the entrenched economic interests of major developing countries while responding directly to demands for climate justice. It could also be said that a further price of initial agreement was a loose definition of objectives which, among other things, provided ample scope for special pleading on sources and sinks. The decision-making procedures of the regime and the failure to agree voting rules provided veto opportunities for self-interested Parties, alongside the many which already existed within national political systems. Shortly after the final entry into force of the Kyoto Protocol one commentator described the regime as 'ossified' and incapable of learning from experience in the ways that one might expect of a long-established international institution (Depledge, 2006). Since

2007 and the Bali Programme of Action there has been a discernible alteration in norms and principles indicating regime change. Although the 'bifurcation' of the regime in terms of Annex I and non-Annex I remains, even in the institution of co-chairmanships of UNFCCC bodies, there has been movement in the direction of a new comprehensive agreement involving mitigation action by all Parties. Of course the nature of the 'differentiation' between Parties remains hotly contested and, for many, the founding principles of the Convention remain sacrosanct. However, at the same time, there has been a retreat from the mitigation principles of the Kyoto Protocol. The price of agreement on the 2011 Durban Platform by developing countries was an undertaking by the EU that it would, unlike Canada, Japan and others, engage in a Second Commitment Period. But it has also become clear that a 2020 agreement will not resemble Kyoto. Instead negotiators have adopted a looser 'bottom up' approach to collective mitigation efforts that substitutes 'contributions' for 'commitments'.

More positively, the UNFCCC, from its inception, has organised and provided 'capacity-building' funds for a vital international reporting effort without which Parties would not have compiled comparable data and inventories. This was their sole concrete obligation under the Convention. Subsequently, the Kyoto Protocol provided a novel experiment and a painstakingly constructed international architecture for emissions trading and carbon offsets with an innovative compliance system. Technology transfer, funding mechanisms, forestry initiatives for the preservation of sinks and new departures on supporting adaptation in developing countries, alongside compensation for 'loss and damage', have all evolved, if often in embryonic form and at a snail's pace. Finally, the often predicted collapse of the entire system has not occurred, but, as will be discussed in Chapter 6, the UNFCCC also serves a number of functions for its state Parties that may be largely unrelated to the search for an efficient international means to curb climate change.

4
Interests and Alignments

Both realist and liberal interpretations of state behaviour and the possibility of international cooperation are founded upon the notion of interest. In realist conceptions of national interest the survival of the state and its territorial integrity are paramount. Arnold Wolfers (1962) made a key distinction between 'possession' and 'milieu' goals. The pursuit of both, in his view, could serve the national interest, but, as he pointed out, realists have tended to define the national interest in terms of possession goals, typically involving the defence of national territory and economic assets. Milieu goals, as the name suggests, involve the general improvement of the international context through the non-exclusive provision of universal benefits and public goods. Environmental agreements typically serve milieu goals through the avoidance or reduction of harm. Avoiding dangerous anthropogenic climate change would be a clear example of the achievement of a milieu goal. For realists, possession goals lie at the heart of national interest and must, therefore, take priority over milieu goals, however desirable they may be. The primary possession goal for a state will be maintenance of territorial integrity plus the protection and extension of important economic resources and strategically significant positions. The exact nature of such national interests will vary over time and with respect to the specific situation of particular states, but realists often observe long-term historical continuities that can sometimes allow interests to be portrayed as having an objective character.

The point about climate change, in contrast to most other environmental issues, is that interests will often be seen to reside within the 'possession', rather than 'milieu', category. This is more than evident for members of the AOSIS, whose territorial survival is at stake. Core economic interests can also be threatened by international attempts to mitigate GHG emissions. For this reason, Saudi Arabia has attempted to block measures that might restrict the use of fossil fuels and compromise the future of its economy. Russia has defended its allocation of tradable 'hot air' under the Kyoto Protocol. China vigorously protects its right to emit those increasing quantities of GHGs upon which its continued economic development is seen to depend. This is a matter of high priority for the survival of a regime that sees its future as dependent on the economic growth that will allow the urbanisation of the impoverished rural masses that still make up around half of China's population.

Liberal explanations of international cooperation involve distributive and integrative bargaining processes where the, largely economic, interests of states potentially overlap in ways that can promote absolute rather than relative gains. In fact, the difference between neorealist and liberal accounts of international interaction has sometimes been narrowed down to just such a distinction between a positive sum or zero sum conception of rival state interests (Grieco, 1988). Discussion of environmental negotiations has centred on interest-based bargaining. The question of how the external environmental interests of a government will be determined and whether a state will be a 'leader or laggard' has been systematically portrayed as a balance between the economic costs of abatement and perceptions of environmental vulnerability (Sprinz and Vaahtoranta, 1994).

This simple dichotomy has provided quite a satisfactory account of the 1985–7 Montreal Protocol negotiations and those for the 1985 Helsinki Protocol on transboundary sulphur deposition.[1] Application of such a parsimonious model to climate change would prove to be much more problematic. Governments and interest groups have been more than willing to assess the costs of mitigation, often in ways that wilfully neglect the benefits of action and the severe cost and risks of inaction, a point made forcibly by the Stern Review (2007). Compared to the acid rain and stratospheric ozone cases, the calculation of national ecological vulnerability is infinitely more complex and subject to continuing dispute and uncertainty. Neat indices of vulnerability are not available for the manifold effects of climatic alteration and the same is true for the cost estimates of adaptation policies. There are indicators of climate risk based on observed events, but these fail to provide an indication of what might be in store for the immediate future. A critical point is that the level of actual vulnerability and damage will more probably be determined by the level of socioeconomic development and adaptive capacity of individual communities and societies, than by the magnitude of climatic change itself. Social and economic changes leading to increasingly exposed human habitations and infrastructure are key drivers of projected increases in climate change impacts and will particularly affect areas with low adaptive capacity which will, in turn, tend to deepen existing social inequities (European Environment Agency, 2012, p. 25). Nevertheless, the idea of a balance between costs and vulnerabilities provides a useful starting point for the identification of national interests and alignments. It helps to clarify what are often inarticulate assumptions in analyses of climate politics.

National interest relates to the assumption that the relevant actors are sovereign nation-states. For the climate change negotiations this, for the most part, reflects the formal situation with the state Parties. There is one evident exception – the EU. The Union alongside its constituent 28 member states is formally recognised as a participant Regional Economic Integration Organization (REIO). The notion of a single national interest has always been something of a fiction. Behind it lurk manifold separate and often conflicting interests, and the positions taken up by governments are frequently the result of extensive bureaucratic politics and lobbying. Industrial and commercial interest groups and NGOs have been a significant part of the climate politics scene and are often treated as actors in their own right alongside state governments. Robert Falkner (2008) demonstrates how the various business coalitions impacted the climate negotiations. In the beginning, the fossil fuel industries, transnational oil corporations and manufacturing allies set up the Global Climate Coalition to counter the IPCC through the encouragement of a sceptical attitude towards climate science and to emphasise, particularly to the US government, the potential economic costs of emissions mitigation. In this they were relatively successful. Their formal representations and lobbying clearly helped to mould conceptions of national interest among the OECD countries, but with the United States they were 'pushing at an open door', because of the existing anti-regulatory stance of Republican administrations and much of Congress.

Similar efforts were rather less successful, however, on the other side of the Atlantic. Fierce industrial lobbying over a carbon tax led to its abandonment but not to any reduction of EU aspirations to take the lead on international emissions reduction. Business interests have not been monolithic. As Falkner (2008, p. 99) says 'Whether climate change presents itself as a threat or opportunity to corporations is by far the most fundamental dividing line in climate related business strategies'. Thus, by the time of the 1995 Berlin Mandate, clear fissures had emerged that pitted, for example, the insurance industries that faced incalculable losses under 'business as usual' scenarios against the fossil fuel producers. Even the latter were soon to be divided over the future of fossil fuels, as opposed to renewables, and with the negotiation of the Kyoto Protocol there were opportunities for business to exploit the flexibility mechanisms. There was then a persistent call from many corporate interests for governments to put in place long-term climate policies that would provide the necessary signals for future capital investment.

Governments will be open to corporate influence if, as most do, they conceptualise the climate problem in terms of the welfare of their national economy and specific sectors within it. The relationship between national policymaking and representation and the corporate sector (and in some instances scientific and NGO activists) can be intimate. The presidency of George W. Bush after 2001 provides a well-known example, as former employees of energy corporations took up senior positions in the administration. National delegates to climate conferences may, on closer inspection, turn out to be representatives and even employees of narrower corporate or sectoral interests. This is one reason why research into the composition of delegations at climate conferences is significant. The other is that the size and composition of delegations has a direct bearing on the ability of countries, particularly LDCs, to participate effectively in negotiations.

Much of the evidence on delegation composition and corporate influence tends to be anecdotal, but Schroeder et al. (2012) have provided some systematic data from COPs. This paints a diverse picture of delegation composition. Brazil, for example, tends to favour representatives of business associations, Russia science and academia and the US legislators. This leads to the conclusion that 'the climate change issue and its associated interests are framed quite differently across countries' (ibid., pp. 835–6). Delegation composition also varies in terms of the representation of government departments, with a trend towards the inclusion of a wider range of ministries. The available data does not support an assertion that corporate interests have 'captured' national delegations, although they may already have strong links with government departments, the European Commission, or congressional or parliamentary representatives. The one finding that appears with the utmost clarity is the relatively small and often tiny delegation size of most developing countries. As has been observed elsewhere, this is indicative of a major structural disadvantage for developing countries which, with limited resources in terms of expertise and finance, struggle to keep abreast of negotiating processes that have become more complex and disparate. Roberts and Parks (2007, p. 14) describe it tellingly, as the equivalent of 'one man against an army'.

In international relations it is to be expected that national interests will be aggregated in alliances or negotiating coalitions. The major alignments in climate diplomacy are outlined below, along with a brief discussion of the interests of their members. They include the G77 with its various subgroups including AOSIS and BASIC and the Umbrella Group of developed world Annex I states. The EU represents the other major negotiating bloc, but it is difficult to categorise. Comprised of 28 member states with differing levels of economic development and vulnerability to shifts in the patterns of international energy use, it too could be described as a coalition of divergent climate-related interests. Yet the EU, with its unique internal structures and policies, is much more than that.

The European Union

In global environmental negotiations 'Europe' comprises both the member states and the Commission operating under shared policy competence. These rather arcane arrangements depend on treaty provisions and case law that determine the extent to which decision-making authority has passed from the member states to the Union. In trade and in some areas of environmental policy this transfer is complete and exclusive. The Commission will, under exclusive competence, have the right to initiate policy and negotiate externally on a mandate agreed by the Council of Ministers. This is not the case for climate policy. It was originally seen as a matter of environmental policy, where the EU had already acquired significant competences, but there were also continuing national competences for the highly relevant areas of energy and taxation. This has meant that competence is shared between member states and the Union, effectively the Commission, and sometimes subject to dispute.[2] There have been occasional displays of disunity, but strangely enough these arrangements have not resulted in the Union being mired in endless coordination meetings and incapable of effective negotiation.

In climate negotiations the Union is represented by the member state holding the rotating presidency of the Council of Ministers and the Commission who sit side by side behind the EU 'plate'. The representatives of member states will also be present and it is sometimes said that the EU negotiates 'at 29' (28 member states and the Commission). Because of the complexity of climate issues, a procedure has been established where certain 'lead states' have a continuing role in developing and representing the EU's position (Oberthür and Roche Kelly, 2007).

In the climate change Convention, the EU has the unique status of a REIO that allows it to participate and sign treaties alongside its member states, according to their respective competences. The EU also heads a group of associated states that have been persuaded, usually on the basis of their membership aspirations, to support the Union's positions. In a submission on the 2015 climate agreement they included Albania, Bosnia-Herzegovina, Iceland, FYROM, Montenegro and Serbia (European Union, 2013). Norway and Turkey have also been associated with EU positions.

The Union, comprising 28 diverse member states at various levels of economic development, has to accommodate a complex set of national interests in its external climate policy. France relies heavily on nuclear power, thereby minimising GHG emissions, while Poland mines and burns large quantities of coal. For Germany, which has dispensed with nuclear power generation and for the eastern members of the Union, there is a continuing high level of dependence on imported Russian gas. States in the South and East of Europe with less-developed economies, can also make legitimate demands for actual increases in their right to burn hydrocarbons in order to ensure the 'cohesion' of the Union. The extent of vulnerability to climate change impacts also varies widely and are only the beginnings of adaptation policy at Union and national levels. Predictions include increased flooding in Northern Europe, declining fisheries and the severe impact of temperature rises in Mediterranean regions. Yet they remain impossible to quantify with any accuracy (European Environment Agency, 2012).

Climate action differs from many other aspects of the EU's foreign relations, because of its intricate connection to the internal energy policies of the Union. As Oberthür and Pallemaerts (2010, p. 27) observe: 'Throughout their two decades of history, international and European climate policy have evolved in tandem and have fed back on each other'. The EU has struggled, since its inception, to produce an effective common energy policy and internal market, while, at the same time, leading the way in climate policy and emissions trading with ambitious plans for energy security through decarbonisation (Vogler, 2013).

Originally, despite the multi-sectoral ramifications of climate change, the Union chose to classify it in terms of environmental policy, allowing the lead to be taken by the Commission's Delegation General (DG) Environment and the Environmental formation of the Council of Ministers. Between 2010 and 2014 there was a dedicated DG for Climate Action which was, in turn, replaced by a new DG bringing together

Climate Action and Energy, under a vice-president responsible for the creation of an Energy Union.

The EU was an early leader in climate politics, sponsoring a 'targets and timetables' approach and embracing an acceptance of the needs of developing countries. The European Council took the probable consequences of climate change seriously, adopting the 2 °C threshold as early as 1996. As will be discussed in Chapter 5, it was allowed to take up advanced positions, without suffering difficult internal economic costs, because of the peculiarly favourable 1990 baseline that allowed negotiation of the pre-Kyoto 'burden sharing agreement' between the member states, to achieve the required 8 per cent collective emissions reduction.[3]

With the negotiation of the Kyoto Protocol, emissions trading was rapidly adopted. Although running counter to the Union's regulatory tradition of 'command and control', the new Emissions Trading Scheme (ETS), introduced in 2005, became its flagship policy (Cass, 2005). As will be recounted in Chapter 6, the EU positioned itself at the heart of climate policy and in direct opposition to the negative policies of the Bush Administration after 2001. In advance of the Copenhagen meeting, the EU made considerable internal efforts to produce a new climate and energy package that would give credibility to its proposed emissions targets. This required, for the first time, internal policy changes that would entail significant costs for member states and industrial sectors. The year 2008 was marked by widespread banking failures and the beginning of a long economic downturn. It was against this sombre background that the climate and energy package wound its tortuous path through the processes of co-decision in the European Parliament and at the Council. There were, inevitably, worries that the measures, by raising energy costs, would make the economic situation worse and destroy international competitiveness. Notable was the danger of 'carbon leakage' – the flight of energy-intensive industries such as steel, cement and aluminium production to China and elsewhere. Added to this were the continuing demands by Poland, Italy, Romania and Bulgaria for more equitable treatment under the revised ETS. At a difficult European Council, held in December 2008, the French presidency negotiated a series of compromises on redistribution of allowances, restriction of auctioning and the further use of offset mechanisms, which managed to accommodate the various national economic interests involved. This was a substantial achievement, but it was not to have the desired galvanizing effect on the 2009 Copenhagen climate conference.

The EU position on a 2015 agreement, to be implemented in 2020, stressed the need for a new Protocol under the Convention, which was 'ambitious, legally binding, multilateral, rules-based with global participation and informed by science' (European Union, 2013a). Of all the major Parties the EU was most insistent on establishing timely and verifiable national emissions pledges sufficient to provide certainty and mutual confidence in the achievement of an ambitious agreement. After the usual internal negotiations and concessions to national energy interests the October 2014 European Council was able to announce its Conclusions on a 2030 Climate and Energy Policy Framework (European Council, 2014; Keating, 2014c). Included was a binding 2030 overall target of a 40 per cent reduction in GHG emissions, against a 1990 deadline, as the Union's 'intended contribution'. Less impressive were 'non-binding' targets of 27 per cent increases in energy efficiency and the share of renewables over the same period.

The Umbrella Group

In the early years of the climate regime, developed countries other than EU member states formed a negotiating coalition known by the acronym JUSSCANZ (Japan, US, Switzerland, Canada, Australia and New Zealand). Taking their lead from the United States, they adopted a more restrictive attitude to emissions reductions than the Union and were critical of the allowances made for non-Annex I countries. It is an alignment encountered elsewhere in the UN system, for example on human rights issues, and still meets occasionally. It was active during the negotiations for the Kyoto Protocol, but subsequently metamorphosed into a broader 'Umbrella Group', excluding Switzerland, but including Norway, Russia and Ukraine. The immediate motivation was to avoid restrictions on the use of the flexibility mechanisms, but its remit has widened to include reporting and LULUCF (Yamin and Depledge, 2004, p. 45). As might be expected from the very different national interests and political orientations of its members, it is only a loose and shifting alignment and does not attempt formal coordination of positions and statements. One of the achievements of the EU during the Kyoto ratification process of Kyoto was to split the Umbrella Group by prising key members away from the United States. The position that unites its members is that advanced developing countries should be treated in the same way as the established major emitters within the 'evolving context' of the Convention and their 'respective capabilities'.

United States

The US national interest has been generally conceived of as limiting the impact of any international agreement on its domestic economy and avoiding any legal wording that might imply obligations. During the INC negotiations, the Bush (senior) administration made it clear that, as the then largest emitter of GHGs, the United States was not prepared, during an election year, to agree to an emissions stabilisation target that might damage the American economy. This was coupled with a long-standing position, reflected in the Byrd Hagel Resolution of 1997, that, if the United States were to make reductions commitments, economic competitors and notably China should be required to do likewise.[4] There have been clear differences between Republican and Democratic administrations on climate change. The Clinton administration was prepared to ratify the Convention and then to undertake the negotiation of the Kyoto Protocol (Harris, 2000). The *quid pro quo* for a US commitment to a 7 per cent reduction in emissions was the incorporation of the flexibility mechanisms, based on Californian experiments with sulphur emissions trading and with a monitoring and compliance system that owed much to US experience with strategic arms limitation. Flexibility was intended to avoid direct costs to the US economy, but it was soon clear that achieving the US Kyoto target, based on a 1990 baseline, would have been a great deal more costly than the EU's parallel 8 per cent commitment.

In 2001, the incoming administration of George W. Bush, staffed by numerous former employees of the fossil fuel industry, said that it would renounce the US signature of the Protocol. American policy turned to resolute opposition during the ratification process. At COPs 8 and 9 the United States even espoused a reduction in carbon intensity approach, as a means of finding common ground with China and the G77, whereby all might avoid economically damaging emissions limitation commitments (Roberts and Parks, 2007, p. 144). Opposition to international 'targets and timetables' was coupled with what amounted to official rejection of climate science. Prior to the Gleneagles G8 summit in 2005, 'sherpas' were closeted in Whitehall with their US counterparts, attempting to persuade them not to excise wording to the effect that 'we know climate change is occurring' from the communique.[5] Later in the same year, official perceptions of US invulnerability remained, even in the face of the damage done to New Orleans by Hurricane Katrina. Attempts were made to bypass the whole UNFCCC process, for example with the APEC initiative of 2005 or the Major Economies Forum, together with various technological initiatives that played well to a 'promethean' belief that technology would provide the necessary solutions. The United States was finally persuaded to re-engage with the UNFCCC at Bali in 2007, but only on the basis that it would have no part in the discussion of Kyoto.

The arrival of the Obama presidency raised expectations that the United States would no longer perceive its economic national interest to be in outright opposition to internationally agreed emissions limits. Indeed there was some movement towards an internal decarbonisation policy. As now only the second largest emitter, the United States posted a voluntary pledge of a 17 per cent reduction in emissions by 2020 (against a 2005 baseline), with a projected downward pathway to an 83 per cent reduction by 2050. This was in the context of the shale gas revolution, which, through the widespread adoption of hydraulic fracturing technology, lessened dependence on coal and held out the prospect that the United States would become a net energy exporter. President Obama decisively changed his predecessor's attitude to climate science and vulnerability, declaring in his 2013 State of the Union Address that 'climate change was a fact'. There was furthermore 'no time for a meeting of the flat earth society' when there were moral obligations to future generations. The administration expressed its intention to lead international climate negotiations and set out, in its June 2013 climate policy announcement, a raft of domestic policy changes on standards and modernisation of energy networks, as well as measures for the protection of sinks (United States Executive Office of the President, 2013). All shared a notable characteristic, that they could be implemented through the Environmental Protection Agency (EPA) under existing legislation without resort to a divided Congress. On the occasion of a November 2014 APEC summit in Beijing, President Obama made a joint announcement with the Chinese president proclaiming a new US national target of a net cut in GHG emissions of 26–28 per cent below 2005 levels by 2025, 'an ambitious target grounded in intensive analysis of cost-effective carbon pollution reduction achievable under existing law' (United States Government, 2014).

Japan, Canada, Australia and New Zealand

These Annex I countries accepted commitments under the first period of Kyoto and ratified the Protocol. They are all high-income economies with very substantial per capita emissions. Japan ranks as the third largest national economy on earth and fifth largest GHG emitter and is one of those countries whose participation in a future climate agreement might be regarded as essential – as it was under the terms of ratification of Kyoto. Japan's international climate activism in this period is considered in Chapter 6. Australia and Canada are both major energy exporters, with the latter controversially involved in the development of what is estimated to be the world's third largest oil reserve in the Alberta tar sands. The process of extracting this heavy bituminous crude is judged to be particularly damaging to the environment and Canada has been involved in a dispute with the EU, which has attempted to discriminate against its import by way of its Fuel Quality Directive. Australia is a significant exporter of coal to fuel Chinese industry. By contrast, Japan had few domestic energy resources and was heavily dependent on nuclear power generation. Prior to the March 2011 Tsunami and the meltdown of the Fukushima nuclear reactors, 26 per cent of its power requirement was provided by nuclear energy, a proportion that was planned to increase to 50 per cent (Meltzer, 2011, p. 1). New Zealand is unique, among developed countries, in its emissions profile. It ranks 5th out of 27 OECD countries in terms of per capita emissions, but over 47 per cent of these emissions are from agriculturally related methane and nitrous oxide (United Nations Department of Social and Economic Affairs, 2010).

These are all highly developed and well-educated societies, where there are known vulnerabilities to climate change – Australia, for example, has experienced serious drought. Their governments had a perceived interest in avoiding damage to their economic prospects through excessive commitments to GHG reduction. Nevertheless, for a combination of reasons, including the availability of 'flexibility mechanisms' and extensive diplomatic pressure from the EU, they were able, unlike the United States, to adopt first commitment period obligations under Kyoto (Japan 6 per cent, Canada 6 per cent, Australia 5 per cent). Australia took until late 2007 and the election of a Labour government to ratify the Protocol. The previous Liberal government had claimed that ratification would be 'counter to the nation's interest' (Talberg et al., 2013). Both Australia and New Zealand subsequently developed domestic emissions trading schemes and Japan was extensively involved in CDM schemes.

The flexibility mechanisms did not yield all the benefits that had been expected. Japan became disillusioned with the CDM and from 2010 began to experiment with its own bilateral programmes, the Bilateral Offset Crediting Mechanism (BOCM), which is unrecognised by the UNFCCC. The Abbot government, which came to power in Australia in September 2013, moved immediately to close down its emissions trading scheme. Canada exercised its formal right to withdraw from the Protocol in December 2011, having failed by a large margin to achieve its 6 per cent reduction target and wishing to avoid what were described as 'enormous financial penalties' (Nikiforuk, 2013). In the post-2012 discussions, Japan, Canada and New Zealand all indicated that they would not be prepared to join the EU in participating in a Second Commitment period. Australia agreed to enter into the Second Commitment period at the Doha COP 2012. This appears to have been contingent on a deal with the EU that the Australian domestic emissions trading scheme would be linked to the ETS (*Australian*, 2014). On the arrival of the new Australian Liberal-led government in 2013 and despite the discontinuation of emissions trading, participation in the Second Commitment period remained.

Emissions reduction pledges of varying ambition were posted under the Copenhagen Accord/Cancun Agreement. Australia pledged a 5 per cent reduction against 2000 levels by 2020, rising to 25 per cent conditional on action by other Parties. The Japanese pledge was an ambitious 25 per cent by 2020 against a 1990 baseline, but a dramatic reversal was soon to occur following the Fukushima disaster. At the 2011 COP Japan was notably absent from the coalition that crafted the Durban Platform (Asuka, 2014, p. 27). There appeared to have been a 'massive degradation of ambition', in which the emissions reduction target was scaled back to what would amount to a 3.1 per cent *increase* in 2020 against a 1990 baseline, violating not only the 2020 pledge, but also Japan's original Kyoto commitment (Jeffery et al., 2013). Such a reversal cannot be accounted for simply by the loss of nuclear-generating capacity; it also reflected lobbying by industrialists for the removal of what they regarded as obstacles to international competitiveness (Asuka, 2014, pp. 27–9).

After 2008 the policies and associated conceptions of national interest of all four countries shifted quite dramatically. There are some special reasons in the case of Japan, but overall these Umbrella Group members appear to have re-calculated their national interest to give absolute priority to economic growth, in all probability as a response to the financial crisis that overtook the global economy from 2008 and put pressure on existing governments, resulting in the election of right-wing successors. In was also evident that discussion of the vulnerability to and costs of climate change had been significantly distorted by challenges to climate science and their espousal by right-wing press and politicians. This was reflected by the closing down of climate science and advisory institutions in both Canada and Australia.[6]

In all four countries economic growth was prioritised, sometimes explicitly, at the expense of climate policy. For example, in 2014,

Australian prime minister Abbot, acting as host of a G20 summit, said that he did not wish the agenda to be 'cluttered' by subjects such as climate change 'that would take the focus from his top priority of economic growth' (*Australian*, 2014). The November G20 meeting actually demonstrated the shift in US policy as President Obama insisted on the inclusion of references to an agreement under the UNFCCC.[7]

Russia and economies in transition

The economies of the old Soviet bloc were recognised in the UNFCCC as 'economies in transition'. They were included in Annex I, but exempted from the capacity-funding responsibilities of other Parties. For them the 1990 baseline for emissions reduction calculations was crucial, because it encompassed the extraordinarily high levels of pollution typical of Soviet era industrialisation. Soon economic retrenchment and reorganisation were to bring massive emissions reductions, to the extent that internationally agreed targets could be more than met without any action to restrict emissions. In the Russian case, 2012 emissions were 34.1 per cent below those of 1990, well below its Copenhagen pledge of a 25 per cent reduction by 2020 (European Parliament, 2013, p. 25). A 1990 baseline yielded substantial quantities of 'hot air', which are tradable under the Kyoto Protocol. Russia was also in a favourable position to exploit the ratification procedures of the Protocol. As the fifth largest emitter of GHGs, it was able to drive a hard bargain with the EU as the price of ratification (where its ratification was essential if the Protocol was to enter into force) to procure an agreement from which it would clearly profit. President Putin was, thus, able to secure EU backing for Russian WTO membership and further national benefits – in terms of its ability to count its forests as carbon sinks (Falkner, 2008, p. 131). The Russian government has refused to take on commitments for the second period of Kyoto, but still expected to be able to continue to trade its remaining stocks of 'hot air'. When this was refused at the Doha COP in 2012, it was the occasion of a Russian challenge over Convention procedures. Most of the other 'economies in transition' in Eastern Europe acceded to the EU in 2004. Belarus and Ukraine have tended to align themselves with Russia on climate issues, although both have accepted commitments under the Kyoto Second Commitment period. They have also reaped substantial benefits under the Joint Implementation mechanism of the Protocol.

Russia has the world's largest proven reserves of natural gas and significant oil resources, and is beginning to claim and exploit arctic hydrocarbons. Energy exports have provided the basis of Russian economic recovery after the fall of the Soviet Union and, as was demonstrated in the 2006, 2009 and 2014 gas supply crises, a number of European states are critically dependent on Russian gas exports. Control of energy supplies has become a significant instrument for Russian foreign policy. Therefore, Russian national interest in profiting from its reserves of 'hot air' is balanced by concerns, similar to those of other energy exporters, over a climate agreement that might restrict demand for hydrocarbons. The Ministry of Economic Development stated in 2013 that other countries' environmental policies were more of a concern than its own, because they could have a serious effect on hydrocarbon exports (European Parliament, 2013, p. 5).

At the time of the Kyoto ratification, there were serious debates as to whether Russia would be a beneficiary of climate change, with claims of an increase in the temperate zone available for agriculture ranged against warnings of the consequences of melting permafrost and the dieback of Siberian forests. There was officially endorsed academic scepticism as to the validity of IPCC estimates. Independent surveys suggest that there is now little question that 'Russia is expected to experience some of the earliest and most dramatic effects of climate change – almost all of them bad' (IISA, 2011, p. 1). This understanding appears to be almost entirely absent from Russian policymaking. National economic benefits are in no way counterbalanced by perceptions of environmental vulnerability, and critical environmental voices are muted (Kokorin and Korppoo, 2013). A study of Russian energy and climate policies concludes as follows:

> In all negotiations, Russian positions have been driven more by economic and geostrategic motives than by ecological considerations. Climate change is often only considered for its economic ramifications: either as a threat to the national economy or as an opportunity (e.g. for agriculture and access to new raw materials).
> (European Parliament, 2013, p. 25)

The Environmental Integrity Group (EIG)

Formed in 2000, this is really a meeting of outsiders. Switzerland defected from the Umbrella Group after Kyoto and was joined by non-Annex I OECD members South Korea and Mexico, along with Liechtenstein. These 'strange bedfellows' shared a sense of exclusion in the way in which negotiations and key informal sessions tended to be structured around the major negotiating groups (Yamin and Depledge, 2004, p. 47).

South Korea has experienced spectacular economic growth leading to a doubling of GHG emissions since 1990 (United Nations Department of Economic and Social Affairs, 2010). It is now a major manufacturing power in the global economy. This has not been achieved at the expense of environmental policy and an emissions reduction of 30 per cent against business as usual by 2020 has been pledged. The energy intensity of its production is still very high, but it has put in place a policy package described as 'innovative and exceptional for a non-Annex I country' (Climateactiontracker.org, 2013). Mexico represents another 'progressive advanced' non-Annex I country with well-developed internal climate legislation and an external pledge, which has been increased from 20 per cent to 30 per cent reduction, against a 2020 business as usual reference point (European Parliament, 2013a, p. 82). Switzerland has joined the Second Commitment period of the Kyoto Protocol, with a target of 15.8 per cent reduction against 1990 emissions levels. It has also made a Copenhagen/Cancun pledge equivalent to that of the EU.

G77/China

This broad, UN-based coalition has existed since the formation of structured negotiating groups for the United Nations Conference on Trade and Development (UNCTAD) in 1964. During the 1970s it provided the framework for the campaign for a New International Economic Order and for the Common Heritage Provisions of the Law of the Sea Convention. Its 133 members will, if united, form a majority at the General Assembly, or a UN universal membership Convention such as the UNFCCC.

The interests of members have increasingly tended to diverge, as economic globalisation has widened the disparities between them. They are held together by a common resentment of the colonial past, and of the developed OECD countries, and by extensive institutionalised cooperation. On climate matters, different countries have taken the lead, India coordinating post-Kyoto positions, while Brazil has led on forests. Common positions must be approved by the main group, which gives mandates to lead spokespersons, who are required to report back on negotiations. The G77 chair for negotiations holds a significant position in balancing the various factions and interests, and some have been more successful than others. Yamin and Depledge (2004, p. 37) note the conciliatory and balanced chairmanship of Tanzania, which held the G77 together at Kyoto, and the outstanding diplomatic skills of the Iranian Chair in 2001 that 'contributed greatly' to the constructive participation of the G77 at the Marrakesh COP. At other times, they have been less successful. The Sudanese G77 spokesperson at Copenhagen provoked outrage when he described the Accord, negotiated by prominent members, as 'an incineration pact', based on values that had 'tunnelled six million people in Europe into furnaces' (ENB, 2009, p. 8).

The Alliance of Small Island States (AOSIS)

Formed in 1990, the alliance has 43 members, mainly small island developing states, but also Singapore, Cuba, Malta and Cyprus. Most alliance members make virtually no contribution to the drivers of global climate change, but are at the same time uniquely and visibly vulnerable to its effects, in terms of sea-level rise. It is the intense existential quality of their core national interest in surviving climatic change that lies at the heart of their role. As various commentators have observed (Yamin and Depledge, 2004; Dimitrov, 2010, p. 805; Brenton, 2013), they have been extraordinarily proactive in climate negotiations in ways that are out of all proportion to their tiny size. For example, they produced the first draft of the Kyoto Protocol, have championed the 1.5 °C target and are granted a special seat in the UNFCCC Conference bureau alongside the usual UN regional groups. They have achieved this by the force and persistence of their arguments, strongly supported by the NGO community, and through a range of astute organisational and other strategies that Betzold's (2010) detailed study has described as 'borrowing power' from external sources. A primary element has been AOSIS's normative appeal in relation to concepts of climate justice – considered in Chapter 5.

The least-developed countries

This is a group of countries, identified as such within the UN system, which defines them in terms of 'low income, weak human assets and high economic vulnerability'. Within the UNFCCC the LDC group has 48 members. Five small island Pacific developing states are also members of AOSIS, but the bulk of the group (33 states) are from sub-Saharan Africa. The LDCs are recognised as a special case under Article 4.9 of the Convention. Alongside the small island developing states (SIDS), they are 'among the most vulnerable to extreme weather events and the adverse effects of climate change', having the least capacity to cope with and adapt to climate change (UNFCCC, 2009, p. 4). Accordingly, they have been given technical support and capacity building, along with special assistance in drawing up their National Adaptation Plans, and there is also a dedicated fund, the LDCF. They have a particular interest and involvement with the developing arrangements for 'loss and damage' to assist with disaster management and to deal with problems of inadequate insurance. In common with the rest of the developing countries they will have an important national interest in ensuring that adaptation funding is available through the climate regime and that the promises made by Annex I countries are adhered to. All this might suggest that the involvement of LDC members is limited to winning adaptation funding and compensation. This is not the case for LDCs have involved themselves in mitigation actions through their national development plans (Kirby, 2013).

The Organization of Petroleum Exporting Countries (OPEC)

Saudi Arabia, as the leading member of OPEC, played a consistently negative role in the development of the Convention and Protocol, skilfully exploiting its position within the G77 and its privileged relations with both developed countries and transnational oil companies (Depledge, 2008). It was largely responsible for Article 4.8 of the Convention and 2.3 and 3.14 of the Protocol, which call for full consideration of the needs of countries whose economies are highly dependent on incomes from fossil fuels and, of course, for the deadlock over voting procedures. Decarbonisation might appear to threaten the very basis of some OPEC economies and the revenue streams of others. However, OPEC is not a homogenous bloc in terms of the composition of its members' economies, or in terms of the very wide variation in per capita emissions. It includes developing countries such as Nigeria, Indonesia and Venezuela, with a range of other interests and alignments. OPEC members are also likely to suffer the adverse effects of climate change and have pressed for international assistance with the diversification of their economies, mitigation activities (including an end to flaring) and fuel switching to natural gas, plus adaptation (Karas and Bosteels, 2005). Their demand, in advance of Kyoto, was for 'compensation' to fund diversification of their economies. A 2013 OPEC statement stressed the need for 'full consensus' on a future agreement which should 'minimize any adverse impacts and assist OPEC members and other developing countries to adapt by diversifying their economies, strengthening their resilience and enhancing increased investment and technology transfer' (OPEC, 2013).

BASIC

The BASIC coalition has risen to play a central role in climate negotiations in recent years. Comprising Brazil, South Africa, India and China, it was formed prior to the 2009 Copenhagen COP, becoming the direct interlocutor of the United States in the drafting of the Accord. It has institutionalised itself with regular quarterly ministerial meetings since then. A view of its operations is provided in a leaked US diplomatic cable:

> It is remarkable how closely co-ordinated the BASIC group has become in international fora, taking turns to impede USA/EU initiatives and playing the US and EU off against each other. BASIC countries have widely differing interests, but have subordinated these to their common short term goals. The US and EU need to learn from this coordination and work more closely and effectively together ourselves.
>
> (US Embassy Cable, 3 February 2010)

What are described as 'common short term goals' actually rest upon a set of positions that do not require additional actions by BASIC members and emphasise their continuing solidarity with the rest of the G77. A Second Commitment period for the Kyoto Protocol and full recognition of developed country responsibilities, together with implementation of their mitigation and funding and adaptation commitments, have been standard requirements, alongside the observation that the voluntary actions of developing countries 'constitute far greater in quantum and impact than those of developed countries' (BASIC, 2013, p. 7). The principles of the UNFCCC are sacrosanct and unilateral actions, such as those taken by the EU on aviation, are unacceptable departures from 'multilateralism'. Above all, any new climate agreement must be in line with 'common but differentiated responsibilities' and the division between Annex I and the rest. The Durban Platform was, therefore, 'by no means a process to negotiate a new regime, nor to renegotiate, rewrite or re-interpret the Convention and its principles and provisions' (ibid., p. 12).

People's Republic of China (PRC)

China has risen to challenge for the position of the largest economy on earth, with a GDP per capita that begins to approach those of Annex I countries and per capita carbon emissions that, in 2013, exceeded those of the EU.[8] With major cities at sea level, Chinese policymakers are acutely aware of the risks of continuing climate change, but their overall priority is to ensure the stability of the regime through continuing rapid industrialisation and economic growth. Only in this way can the requirement to urbanise most of the remaining half of the Chinese population that continues to languish in rural poverty be achievable. It is a massive task, requiring huge inputs of energy. China is the world's leading coal producer, its output being almost four times that of its nearest rival the US (World Coal Association, 2015). While the commissioning of new coal-fired generating capacity is often remarked upon, China's coal consumption, that currently meets 70 per cent of its primary energy needs, is planned to level off at around 2020 levels. The serious air pollution of Chinese cities is now a pressing policy concern and the PRC government has been engaged with the provision of hydro, solar and other types of renewable energy, on a scale that is unmatched elsewhere.

Originally, China's economic national interest 'motivated participation in the climate change regime with low cost commitments', based on *no regrets* internal policies that might receive external support (Zhang, 2003, p. 82). External obligations detrimental to economic growth would not be undertaken. There was some modification to this stance when, prior to Copenhagen, China was prepared, for the first time, to reflect its internal drive for energy efficiency in an international pledge to reduce the energy carbon intensity of each unit of GDP by 40–45 per cent from 2005 levels by 2020. China has also been extensively involved with bilateral technology schemes with the EU, US and other developed countries and has been the largest user of CDM projects, which has given it an economic stake in the climate regime (Dai and Diao, 2011). It is also experimenting with the introduction of emissions trading. However, as Christoff (2010, p. 645) comments, China is 'caught in a Faustian policy trap' needing 'ongoing domestic growth of around eight per cent per annum to maintain social and political stability'. Yet the environmental and social cost of growth based on continued reliance on cheap coal could render its pursuit of development ultimately self-defeating.

At Copenhagen, in 2009, a rigid negotiating stance appeared designed to maintain maximum economic flexibility and room for manoeuvre. But in the ensuing years its stance as a developing G77 member increasingly lacked credibility. The assumption of a certain joint responsibility with the United States was manifested first in the 2013 agreement on HFC reduction and then, in Xi Jinping's and Obama's joint announcement of new emissions reduction objectives, at Beijing in late 2014.

The Chinese 'intended contribution' to a new climate agreement was to work towards peak national carbon dioxide emissions 'around 2030' with an intention to peak early and to increase the non-fossil fuel share of all energy production to 20 per cent (United States Government, 2014). This appears to indicate a major shift in the Chinese approach that recognises a national obligation to engage with the developed countries in mitigation activities.

India

India is usually bracketed with China as a giant developing economy; and is sometimes seen as a serious rival to an emergent Chinese superpower. As the fourth largest emitter of GHGs, with the potential to go well beyond this, it cannot be ignored in climate politics. However, the differences in terms of the scale and level of economic development are stark. In the view of Hurrell and Sengupta (2012, p. 472), India 'has as much in common with the least developed countries as with other BASIC states'. Indian GDP and emissions per capita – two tonnes of carbon – (Global Carbon Project, 2014) are only a fraction of those of China. It also has major vulnerabilities to climate change, in terms of its water supplies and sea-level rise, not to mention the implications for agriculture in what remains a desperately impoverished and largely rural society.

Economic development and poverty reduction are likely to remain the twin material drivers of Indian climate policy. In a country in which hundreds of millions of people have no access to electricity, there is huge pressure to expand the import and use of hydrocarbons. 'Policies linked to mitigation are generally motivated by material concerns over depleting resources, ambitions for maintaining high macro-economic growth, expanding energy access and increasing energy security' (Atteridge et al., 2012, p. 72). In the provision of alternatives to hydrocarbons there is no comparison to what is occurring in China, and attempts to achieve mitigation are hampered both by the small scale of Indian enterprises and by the federal structure of relatively autonomous state governments. It is, thus, unsurprising that for a long period Indian external climate policy was marked by its distrust of other Parties, refusal to take up any position on Indian mitigation and a general defensiveness, reflected in a punctilious insistence on equity and differentiated responsibilities of the developed world. There was an observable change, in advance of the Copenhagen COP, when for the first time, and in a direct response to the Chinese pledge, India announced a lesser emissions intensity reduction target of 20–25 per cent. Generally, however, India's hard line position in negotiations has been maintained, notably in 2011 when its insistence on weakening language on the legal form of a new agreement provided the last barrier to acceptance of the Durban Platform (ENB, 2011, p. 27).

Brazil

As the seventh largest emitter of GHGs, Brazil presents a contrast. Its energy requirements have been met by reliance on hydroelectric power and non-fossil fuels, with its pattern of national emissions dominated by the effects of deforestation. The latter peaked in 2005 and since then deforestation has been decoupled from economic growth, which has promoted national acceptance of REDD+. On the formation of the BASIC group there was national political and business consensus on the importance of international climate action (Hochstetler and Viola, 2012) and the 2008 economic crisis did no extensive damage to the Brazilian economy. The Copenhagen Accord was favourably received and Brazil made the strongest of all the BASIC pledges, to a 36–38 per cent reduction against business as usual (BAU) by 2020. This is equivalent to stabilisation at 2005 levels and is to be achieved mainly through a reduction in deforestation, agricultural measures and changes in land use (European Parliament, 2013a, p. 80).

Republic of South Africa

South Africa, the final member of the BASIC group, is different again. As the pre-eminent economy in Africa, it has relatively high per capita emissions, equivalent to the EU average, with a profile that 'differs substantially from that of other developing countries at a similar stage of development' (Republic of South Africa, 2011). Some 80 per cent of its GHG emissions derive from inadequate and dangerously insecure coal-fired power generation in an economy with important mining and minerals export and processing sectors. Agricultural and land-use emissions are only around 5 per cent of the total (the developing world average is 44 per cent) (ibid.). However, it accounts for a much smaller share of global GHG emissions than other BASIC members.

Both the private sector and government in South Africa are interested in the commercial opportunities in carbon financing that may be the product of international negotiations, rather than in the strict pursuit of a North–South agenda. Job creation for a large and poor urban population is a major challenge for African National Congress (ANC) governments charged with having failed to deliver on the economic promises surrounding the end of Apartheid rule. Over the medium term, climate change poses a threat to water resources that are 'already over-committed with potentially serious effects upon industry, maize production and the hopes for social improvement and poverty alleviation' (Madzwamuse, 2014). The government's position on mitigation is to offer 34 and 42 per cent downward deviation from emissions levels, projected under business as usual assumptions for 2020 and 2025. In common with other developing country actions, the 'extent to which it can be achieved depends upon the extent to which developed countries meet their commitment to provide finance, capacity building, technology development and transfer' (Republic of South Africa, 2011, p. 25).

ALBA and AILAC

Some of the most intriguing recent developments in climate politics have occurred within Latin America. Prior to its emergence as a key member of the BASIC quartet, Brazil had attempted to exercise a degree of continental leadership, opposed by Argentina and also by a radical grouping of anti-US states. The latter formed ALBA – the Bolivarian Alliance for the Peoples of Our America. Members include Venezuela, Cuba, Bolivia, Nicaragua, Dominica and Ecuador. It was ALBA's intervention that famously blocked the adoption of an agreed text at the 2009 Copenhagen COP, described in detail in Chapter 6. For ALBA members, the climate issue is part of an anti-imperialist struggle, where climate change is a 'consequence of the capitalist system, of the prolonged and unsustainable development of the developed countries (and) of the application and imposition of an absolutely predatory model of development on the rest of the world' (ALBA, 2009). The interests of developing nations are, therefore, seen in terms of resistance, restitution and repayment of the ecological debt of developed countries. In part as a response to events at Copenhagen, where Columbia directly confronted Bolivia, a competing regional alignment, the Association of Independent Latin American and Caribbean States (AILAC) was formed (Chile, Colombia, Costa Rica, Guatemala, Peru) and entered into the climate negotiations as a formal group in 2012. None of these middle-income countries are major emitters, nor do they have particular climate vulnerabilities. They do have to make energy choices (in Chile's case between coal and natural gas) and they all would regard themselves as engaging in progressive domestic decarbonisation and adaptation policies. In what has been described as a 'revolt of the middle', they have argued for immediate action at the international level, wishing to transcend North–South debates and to interpret the principles of the Convention in a dynamic way within a contemporary context (Roberts and Edwards, 2012).

Conclusions

For most countries, considerations of national economic interests and advantages continue to dominate perceptions of vulnerability. In the Anglo-Saxon world, the latter have also been continuously undermined by well-orchestrated campaigns of climate scepticism. The immediate costs of action, in terms of higher energy prices, loss of development potential or international competitiveness, tend to trump the anticipated costs of climate change. Furthermore, the perceived costs of action have been raised by the post-2008 economic downturn and the prioritisation of growth as a matter of political survival. This has been particularly evident for those Annex I countries that reneged upon, or refused to renew, their commitment to the Kyoto Protocol. To put it in terms of Wolfers's (1962) distinction, economic possession goals have far outweighed the common milieu goal of a stable climate. Only relatively rarely has vulnerability to climate change been conceptualised in terms of national possession goals where, as in the case of some AOSIS members, the costs of predicted climate change outweigh everything else and are core matters of national survival. Mainly economic conceptions of national interest have promoted defensive attitudes that impede progress on the climate agenda. Gambia's sometime chair of the LDC group has criticised the way in which negotiations are entrenched, 'These civil servants (of other Parties) work to defend their interests at all costs and so progress towards an effective agreement remains slow' (Jarju, 2014). Since negotiators tend to lack trust in the good faith of other parties and are continually worried that they may be economically disadvantaged, a primary justification for the creation of an international regime is that it gives assurance that competitors will not free ride on the mitigation efforts of others.

However, conceptions of interest are not immutable and national utility functions are not simply a matter of profit and loss. This has been recognised in the international cooperation literature, where economistic concepts of interest have been tempered by an appreciation of the importance of cognitive change for regime formation and the possible impact of 'epistemic communities' (Haas, 1990). Influential commentators, notably Stern (2007), have pointed out that a re-conceptualisation of what is at stake with climate change, and a prudential attitude to future costs and growth, should lead to a re-calculation of national economic interest favourable to immediate action on mitigation. There is evidence that some Parties, notably the EU but also the members of AILAC, are capable of challenging the assumption that economic growth must necessarily come at the expense of higher GHG emissions. They are increasingly aware, too, of the benefits of decarbonisation of their economies.

The obstacles are not merely conceptual. What emerges very clearly from a survey of national positions is that they tend to be the outcome of complex internal and, on occasion transnational, political processes. One of the most profound difficulties in dealing with climate change at the international level is that national policies are conditioned by internal structures, where there can be many veto points for vested interests with huge stakes in the continued use of hydrocarbons. This is most obviously the case with the apparent impossibility of persuading the US Congress to agree to any new legislative (as opposed to executive) action on climate issues. This severely limits and defines the president's scope for international action, however enlightened his or her view is of the national interest and of the validity of IPCC reports.[9] After 2005, the EU has had to pilot its climate policy through the legislative shoals of 'co-decision' between Commission, Council and European Parliament. Even in the ruling Chinese Communist Party, there are indications of difficult domestic struggles over climate policy that have impacted on China's international stance (Christoff, 2010).

Climate alliances and alignments appear to be relatively stable and generally in alignment with levels of economic development and sensitivity to action or inaction within the climate regime. Yet there are also regional and exclusively political factors in play. The United States, for example, aligns with Japan, Canada, Australia and New Zealand. There is close political identification and the United States has defence arrangements with all of them, while they share a set of economic interests and outlooks. However, this is also the case for most of the member states of the EU, which has, throughout, challenged JUSSCANZ and the Umbrella Group (that includes Russia). The latter has fractured at various points in the climate negotiations, notably over ratification of the Kyoto Protocol. On the other side, there is the historic UN-centred alliance of the G77, highly institutionalised and representing what were, at its inception, the newly independent and developing states of the South. It has come under increasing strain as the climate-related interests of its members have diverged. Early on, the small island members of the G77 formed AOSIS, bringing together those with a core national interest in early and extensive international action. On the other wing of the G77, members of OPEC had a distinct interest in ensuring the avoidance of any action that would compromise their principal source of export earnings. In this they could be bracketed with the Russian

Federation. From 2009, the four largest G77 economies formed their own climate bloc, BASIC, although at pains to stress their continuing solidarity with the rest of the G77. Within the G77 there are, also, other regionally defined groups – the Africans, for example, who frequently adopt common statements at Conferences of the Parties. The Association of South East Asian Nations (ASEAN) comprises countries that have been very vulnerable to extreme weather events and has, since 2007, attempted to adopt joint positions on climate policy but with little differentiation from the overall line taken by the G77. They have been distrustful of the UNFCCC and somewhat wary of making mitigation pledges (Goron, 2014, p. 112).

The record does indicate the possibility of aligning the interests of major groups through active diplomacy. The Cartagena Dialogue for Progressive Action provides an important example. With 40 or more diverse national participants, it described itself as 'an informal space, open to countries working together towards an ambitious, comprehensive and legally-binding regime in the UNFCCC and committed domestically to becoming or remaining low carbon economies'. It was formed in the aftermath of 'failure' in Copenhagen, through a joint initiative of the Australian and UK governments. The intent was to promote compromise across the divide between developed and developing countries (Australian Government Department of Climate Change and Energy Efficiency, 2011). It comprises members from all the major negotiating groups, the EU, Umbrella Group (Australia and New Zealand) the BASICs (South Africa), plus members of AILAC Mexico (EIG), Indonesia and the United Arab Emirates. The dialogue appears to have played an important part in promoting the compromises that facilitated the success of the Cancun COP in 2010 (ibid.); and in providing a basis for EU diplomacy in formulating the Durban Platform (van Schaik, 2012).

The Like-Minded Group of Developing Countries (LMDC), set up in 2012, appears to have been a response to such attempts to deconstruct the 'firewall' between Annex I countries and the rest. Leading activists in the LMDC are China and India and it contains ALBA members, the Philippines, Thailand and Saudi Arabia. LMDC is described as 'a platform to exchange views and coordinate positions', but one which is firmly anchored in the G77. Their position is that any new agreement must be under the principles of the Convention. The 'fulcrum of the balance in the Convention lies in Art 4.7, under which the extent to which developing countries implement their commitments ... depends upon the extent to which developed countries implement their commitments to provide finance and technology' (LMDC, 2013). The extent of LMDC's insistence on the maintenance of the Annex I-non-Annex I divide is illustrated by their objection even to the grouping of countries by income level in the 2014 IPCC Fifth Assessment Report (Livemint, 2014).

Pure economic interest-based politics cannot explain everything, although it probably accounts for the major part of national climate change policy positions and is the inevitable starting point of analysis. Subsequent chapters examine other motives. Only the most determined realist would deny the existence of a normative dimension to international politics and climate justice and demands for equity lie at the heart of discussion in the UNFCCC regime, while the pursuit of recognition and status provides important, but intangible, components of a state's national interest.

5
The Pursuit of Justice

> The Parties should protect the climate system for the benefit of present and future generations of humankind on the basis of equity and in accordance with their common but differentiated responsibilities and respective capabilities. Accordingly, the developed countries should take the lead in combating climate change and the adverse effects thereof.
>
> (UNFCCC, Art. 3.1)

The pursuit of justice is inseparable from the politics of climate change. Considering the gross imbalance between the benefits accrued by those high income societies whose emissions triggered the enhanced greenhouse effect and the likely impacts visited upon the poor and vulnerable, who bear little or no responsibility for the problem, it could not be otherwise (Shue, 1995; Elliott, 2006). Climate justice has been defined as the linking of human rights and development 'to achieve a human centred approach, safeguarding the rights of the most vulnerable and sharing the burdens and benefits of climate change and its resolution equitably and fairly' (Mary Robinson Foundation, 2011).

The 2002 Bali Declaration of Principles of Climate Justice, drawn up by a group of prominent NGOs, provides some idea of the range of rights holders, responsibilities, principles and normative injunctions that can enter into climate-related political discourse (India Resource Center, 2003). Rights are attributed especially to communities, indigenous people, women and the poor, but also to 'everyone' and to unborn generations. The content of these rights includes freedom from the adverse effects of climate change, clean air and water, plus compensation and reparation for climate-induced loss and damage. There are also demands for procedural justice, such that communities should be allowed representation in the management of local ecosystems and in national and international processes to address climate change. Responsibilities mainly fall upon governments to educate and to address climate change, but there is also an injunction to individuals that they should minimise their consumption of resources and 're-prioritise' their lifestyles. Principles consistent with climate justice are outlined. Among them is the concept of ecological debt 'that industrialised countries and transnational corporations owe the rest of the world as a result of their appropriation of the planet's capacity to absorb greenhouse gases'. There follows from this the assertion of the strict liability of fossil fuel industries and of the right of victims to compensation and reparation. Solutions are required that do not externalise 'costs to the environment and communities and are in line with the principles of a just transition', while market-based or technological solutions should also 'be subject to principles of democratic accountability, ecological sustainability and social justice'.

This provides some indication of the range of issues and inherent difficulties associated with the pursuit of climate justice. There is an encompassing spatial scope, but also a temporal dimension with GHGs having an atmospheric lifetime of as much as 100 years. Should historic responsibility for the stock of atmospheric carbon be assigned and how is this to be apportioned? How may intergenerational rights of those yet unborn, or societies that have yet to develop, be secured? All this is in the context of an, often extreme, incongruence between the sources of emissions and their impacts.

The UNFCCC, which itself forms part of the evolution of international legal principles, picks out certain rights and duties – often in ambiguous ways. They only partly reflect the broad range of ethical claims that could be made. Article 3.1, which contains the much discussed principle of common but differentiated responsibilities and respective capabilities (CBDR-RC), attempts to provide some basis for allocating benefits and burdens in respect of past actions and present and future generations of humankind. The meaning of 'equity' is left undefined and has remained a source of highly politicised ethical contestation. Yet there is also a clear normative injunction for developed country Parties to 'take the lead', which was subsequently operationalised in the Kyoto Protocol and in the arrangements for capacity building and other climate-related funding. There is also recognition, in Article 3.2, of the disproportionate and asymmetrical impact of climate change on vulnerable developing countries, which 'should be given full consideration'.

More controversial is the 'right' of Parties to sustainable development, which they are also enjoined to 'promote' (Art. 4.4). In the same article it is stated that 'economic development is essential for adopting measures to address climate change'. The subsequent evolution of the Convention has involved COP decisions, at Cancun 2011, that 'Parties should in all climate related actions, fully respect human rights' and, at Doha 2012, on the importance of equal gender representation (1/CP/16 and 23/CP/18).

Ethical theory in international relations

Ideas of justice have been prominent in IR theory, both in terms of the legitimation of war and as the basis for international order.[1] The conventional distinction made in the treatment of international ethics has been between communitarian and cosmopolitan arguments. The realist tradition is often aligned with a communitarian approach.[2] This rejects the possibility of a universal moral order of shared values in favour of a narrow focus on the ethical duty of statespersons towards the survival and advancement of separate national communities. In the tradition of Machiavelli and *realpolitik*, actions are to be judged by their consequences for the state. The ethical point of reference is 'compatriot priority'. There is an assumption that state boundaries, or those of national political communities, 'have so much moral significance that citizens of one state cannot be morally required, even by considerations of elemental fairness, to concern themselves with the welfare of citizens of a different jurisdiction' (Shue, 1999, p. 542). From this narrow and consequentialist perspective the wider ethical debates about the responsibility for and costs and benefits of climate change would be generally disregarded. This might represent the position of some fossil fuel exporting states in climate politics, but even here, as in the case of Saudi claims during the INC negotiations, there is an appeal for fair treatment in respect of economies that are 'vulnerable to the adverse effects of the implementation of measures to respond to climate change' (UNFCCC, Art. 4.10). The specific reference is to 'Parties that are highly dependent on income generated from the production, processing and export and/or consumption of fossil fuels'. They are bracketed (Art. 4.9 9(h)) with other clauses that enjoin parties to give special consideration to disadvantaged developing countries most vulnerable to the effects of climate change.

Subject to the pressures of an increasingly globalised post-Westphalian system, and the particular challenges of coping with transboundary and global environmental change, a narrow communitarianism appears increasingly outmoded (Hurrell, 2007, pp. 216–38). Much closer to trends in international law, and the approach of most states, is the communitarian position of 'pluralism'. This recognises that there are certain minimal state duties and responsibilities that enable co-existence in an 'anarchic society of states'. Associated with English School theory, and writers such as Hedley Bull and John Vincent, the 'pluralist' approach to international ethics not only values the diversity of sovereign political communities, but also comprehends the ethical basis of the practices and rules that ought to be observed in order to preserve them. Over the past century the 'greening of international society has become apparent' and states have accepted a 'basic form of global environmental responsibility' and obligation to participate in multilateral environmental policymaking (Falkner, 2012, p. 514). Of significance to climate change policy is the application of negative, pluralist norms that harm should not be imposed upon other states. This has been reflected in international environmental law and is specifically stated in relation to a state's sovereign right to exploit its own territorial resources – in Principle 21 of the 1972 Stockholm Declaration. The wording is repeated in the preamble to the UNFCCC, which notes the 'responsibility to ensure that activities within their jurisdiction or control do not cause damage to the environment of other states or areas beyond the limits of national jurisdiction'. The latter can be interpreted as a reference to the global atmospheric commons. Here, there is only a negative normative injunction to avoid harm.

Arguments that there can be no positive duties beyond state boundaries are 'especially unpersuasive' when made on behalf of the citizens of wealthy industrialized states in the context of international environmental cooperation (Shue, 1999). In fact, pluralists and OECD state governments do recognise some positive ethical duty of mutual aid when other communities are subject to disasters or adverse environmental effects, or are simply incapable of effective participation in international agreements. There are substantial parts of the climate regime that articulate, and to an extent operationalise, this positive duty to build capacity and render assistance to less-developed and vulnerable countries in their mitigation and adaptation activities. Thus, the duty of mutual climate-related aid has become a generally accepted ethical principle that has been honoured, to a varying extent. However, it need not be, and usually is not, based on any ethical recognition of responsibility for the situation of poor and vulnerable societies faced with climate change. Rather, it rests upon the principle, common to development aid and disaster relief, that it is right to ensure that the conditions of life for human populations in other countries should not be allowed to deteriorate below some basic minimum standards of existence. There is a notion of a certain sort of fairness under conditions of extreme inequality between populations.

The limited scope of a communitarian pluralist approach can appear grossly inadequate under contemporary conditions. Commentators have made this point by direct reference to the truly global and encompassing dimensions of the climate change problem and to the way in which it 'inescapably raises questions of (global) distributive justice' (Caney, 2011, p. 22). Distributive principles are required to determine who is entitled to what level of protection, who should bear the burden of dealing with climate change, the nature of such duties and, finally, the question of rights to emit GHGs. The implication is that only a cosmopolitan stance is relevant.

Cosmopolitanism derives from a philosophical tradition grounded in the European Enlightenment and, in particular, the work of Immanuel Kant. Cosmopolitan thought emphasises the rights of all human beings regardless of their location within national communities.[3] It appears especially relevant to the global atmospheric commons because communities cannot be isolated from the effects of climate change. Political action may be stimulated not only by conceptions of common humanity but, as Andrew Dobson (2005, 2006) argues, through an understanding of relationships of causal responsibility. The GHGs emitted by the myriad activities of modern societies and their absorption by sinks have global impact. The climate regime exemplifies what Andrew Linklater has termed a 'cosmopolitan harm convention', 'to cope with the wholly unique form of harm caused by global warming' (Linklater, 2011, p. 39). Cosmopolitan harm conventions are seen as part of a civilising process in international society, which has moved beyond harm conventions that simply manage war and direct injury (Elliott, 2006).

A final issue between communitarians and cosmopolitans relates to the subject of international climate justice. Communitarians make reference to citizens, but only to those encompassed within a state or national community. Most international climate negotiation centres on the obligations, rights and burdens of states; and therefore reflects a communitarian position. Cosmopolitans refer in a universal sense to individuals and their inherent human rights. Discussion of climate rights and obligations may look very different if it is acknowledged that to speak of rich and poor countries may be highly misleading. There are many rich individuals in poor developing countries, just as there are areas of dire poverty in nations at the top of the global income distribution. This line of thinking points towards an individually egalitarian form of climate justice which allocates all individuals a minimum right to emit GHGs, but also penalises and taxes those whose personal emissions are disproportionately high. Attempts to promote individual human rights beyond national borders have been bedevilled by cultural relativism and by the occasionally disastrous consequences of 'liberal humanitarian interventionism'. Some years ago John Vincent (1986) suggested it might be possible to resolve divisions over legitimate individual rights and the allocation of blame in international society, by generating a consensus on the fundamental human right to avoid starvation, without which all other putative rights were rendered meaningless. It may be that this idea could have application to a minimal right to a subsistence share of global emissions.

Two substantive areas of ethical concern will be considered in the remainder of this chapter, where practical examples of both communitarian and cosmopolitan principles may be discerned. The first is central to the mitigation ambitions of the Convention. It involves an examination of the different models for allocating the burdens of GHG mitigation. There are two key issues of fairness here. First, to what extent should historic responsibility for climate change be recognised? Establishing equity here would involve considering the extent to which the past emissions of developed countries have placed an unfair burden of costs on the developing world that requires redress. Second, if it is acknowledged that managing climate change must of necessity be a common endeavour, then there should be some means of allocating the costs in a fair manner, even to those who can claim that they have little historic responsibility for the problem.

The other area that has assumed much greater importance in recent years concerns adaptation and the question of compensation for climate change-related loss and damage. This raises issues as to the basis and extent of state duties and responsibilities to developing societies and the key question of direct liability and compensation. Argumentations about adaptation and mitigation are intellectually entwined through concepts of responsibility and developmental space but, certainly since the Bali Plan of Action in 2007, they are also interdependent in terms of practical negotiation.

Principles of mitigation

Managing mitigation on a global basis must require some agreement on principles of global distributive justice in terms of the allocation of responsibilities and rights, which will determine norms of behaviour. The UNFCCC acknowledges this in a general way in its agreed principles of CBDR-RC and 'equity'; and in terms of the agreement that Annex I Parties were obligated to make the first move. Beyond that, there has been enormous room for controversy, in which Parties, interest and activist groups have been guided by very different concepts of fairness and responsibility within varying temporal frames of reference. The Indian delegation at COP 19 stated, for example, that equity was an absolute and inalienable right that 'cannot be equated with and is far beyond fairness' (ENB, 2013, p. 27). In another formulation, 'what diplomats call equity incorporates aspects of what ordinary people everywhere call fairness. The concept of fairness is neither Eastern nor Western, Northern nor Southern, but universal' (Shue, 1999, pp. 531–2). The following sections outline four, or perhaps five, different models of fairness that have figured in international discussions of mitigation.

Nationally defined contributions

This principle can be discerned in the positions adopted by the United States and other members of the Umbrella Group concerning a 2015 climate agreement. It is close to the communitarian end of the scale in its refusal to acknowledge historic responsibilities or issues of relative economic prosperity. Given our understanding of the long-term cumulative effects of industrialisation and the way that these are disproportionately visited upon poor communities in the developing world, a very powerful argument could be made that the 'no harm' principle should apply not just to specific instances of transboundary pollution, but to high and historic levels of GHG emission. No such acknowledgement has been made by Umbrella Group countries, rather there has been a campaign to excise references to 'common but differentiated responsibilities' from international legal texts. There is, furthermore, an explicit rejection of the notion of equity as a standard to determine national commitments to GHG reduction. As the chief US delegate, Tod Stern, is reputed to have exclaimed during the Doha COP discussions on the Durban Platform for a new climate agreement 'If equity is in, we're out' (Pickering et al., 2013, p. 423).

Instead there is support for 'a structure of nationally determined mitigation commitments, which allow countries to "self-differentiate"' (Stern, 2013, p. 3). 'Contributions should be nationally determined by the Party in question, taking into account the factors it considers relevant' (United States Government, 2014). This means that differing levels of development and national preferences can be accommodated in an international agreement that could be based on a range of possible quantitative and qualitative contributions, including 'hard caps' on emissions and emissions intensity targets. This is acceptable, but hardly optimal, for BASIC countries which have expressed their contributions either in terms of emissions intensity targets or reductions against 'business as usual' projections. Using emissions intensity rather than emissions reduction targets has been an attractive alternative option for some years. Intensity targets were the Bush administration's alternative to the Kyoto Protocol in 2002. As they did not impose any quantified emissions reduction limitation to future economic growth, they became part of the Bush administration's campaign against the Kyoto Protocol and even formed part of an attempt to make common cause with G77 members at COP 9 in New Delhi (Ott, 2002). Carbon intensity approaches to sharing the burden of emissions reduction have the advantage of appearing to demonstrate progress, while encouraging energy efficiency and allowing economic growth. There is, of course, no accounting for existing stocks of carbon produced by an economy and no guarantee that actual future emissions will be reduced as growth continues apace. As Roberts and Parks (2007, p. 142) note, it is possible to make out a Benthamite case for an emissions intensity reduction policy. Utilitarian conceptions of justice would focus on the aggregate net benefits of a policy, including economic welfare. 'The fair solution, with respect to reductions of greenhouse gas emissions, would therefore be to stabilize the climate as effectively as possible while maximizing economic growth.' This, incidentally, would also meet the objections of 'sceptical environmentalists', such as Bjørn Lomborg (2001), who make a cost-benefit case against spending money on GHG mitigation, rather than on a range of other pressing welfare concerns.

Underlying the 'nationally defined contributions' approach is a pluralist notion of fairness which is ahistorical, but in accordance with prevailing principles to be found elsewhere in the international system.[4] GATT/WTO principles are prominent. They emphasise the equal treatment of economies in tariff terms with respect to 'most favoured nation' rules and 'national treatment' of imported goods. There are some exceptions for developing nations eligible for the Generalised System of Preferences. The idea is that fairness requires 'a level playing field' upon which national economies can compete. The oft-reiterated objections of developed world politicians to the application of the CBDR-RC principle rest upon this notion of fair treatment. Its most famous expression is to be found in the 1997 Byrd Hagel Resolution, in which the US government was warned that it should not sign any 'unequal' treaty under which US business would have to compete on unequal terms. Similar worries about competitiveness and the loss of industries and jobs underlie European discussions of the imposition of border tax adjustments for products with a substantial energy component, produced in developing countries that are not subject to emissions trading or carbon taxes.

Targets and timetables

This is the Kyoto Protocol model that contains, at its heart, a conception of equitable treatment that acknowledges, to some extent, the different historical contributions to GHG emissions of Annex I countries and the rest. Recognising the CBDR-RC principle, it operationalises it by designating a range of 'quantified emissions limitation and reduction objectives' (QELROs) for Annex I countries. This acceptance of responsibility was not based on any developed methodology for ascertaining different historic responsibilities for the stock of carbon in the atmosphere, nor did it refer, except in a gross and undifferentiated sense, to the national income of Annex I participants. The reductions agreed under the Protocol were based on 'grandfathered' rights to emit, that is to say the actual emissions levels at the 1990 baseline. This would evidently be an extremely unfair way of designing a global system because, for most developing countries, existing emissions levels are comparatively low. Legitimate aspirations for economic growth would be choked off, while the existing advantages of developed country emitters would be enshrined.

In the negotiation of its internal 'burden-sharing' prior to Kyoto, the EU exemplified, in microcosm, the potential international problem of allocating rights to emit. The EU burden-sharing agreement not only contained major reductions from Germany and the United Kingdom, using a highly fortuitous 1990 baseline, but also allowed very significant increases in emissions from less-developed countries within the Union, including Ireland and the 'cohesion' economies of Southern Europe (Figures 5.1 and 5.2). In the actual Kyoto negotiations there was no systematic attempt to allocate responsibility, rather a diplomatically negotiated set of numbers averaging a 5.2 per cent reduction against a 1990 baseline. The EU had gone into the negotiations expecting that it might have to meet its own ambitious target of 10 per cent or even 15 per cent reductions, which it was able to meet without making any unpredicted emissions reductions (Vogler, 2011) . There was, in the end, no need to employ an innovative 'triptych' formula, developed in the Netherlands for balancing and distributing different types of emissions over time (Ringius, 1997). With a negotiated target of 8 per cent by the

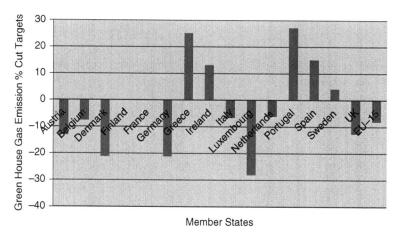

Figure 5.1 The EU burden-sharing agreement 2008
Source: EEA (2006) *'Greenhouse gas emission targets of EU-15 Member States for 2008–2012 relative to base-year emissions under the EU burden-sharing decision'*. Available at: http://www.eea.europa.eu/data-and-maps/figures/greenhouse-gas-emission-targets-of-eu-15-member-states-for-2008-2012-relative-to-base-year-emissions-under-the-eu-burden-sharing-decision Accessed: 28/06/2014.

end of the first commitment period of 2009–12, it was able to revise the burden-sharing agreement downwards.

The Kyoto agreement came to have a symbolic importance for the G77 simply because, despite its inadequacies, it did represent the implementation and continuing validity of a crude version of equity based on CBDR-RC. For the second commitment period, as we have seen, Canada, Japan and Russia were not prepared to subscribe. Some elements of the Kyoto model survive in the EU's approach to the 2015 negotiations which 'must be ambitious, legally binding, multilateral, rules-based with global participation and informed by science ... it should also fully respect the principles of the Convention' (EU, 2013a, p. 1). Kyoto-like targets and timetables are not proposed, but there is the suggestion that something more than a set of national pledges is required. This is apparent in the EU's 'step-wise approach' whereby mitigation commitments would be subject to a review and negotiation process prior to their inscription in an agreement. There should, therefore, be 'a robust international assessment of individual and collective ambition of commitments in light of the below 2°C objective'. This would involve comparison and an assessment of the ambition and

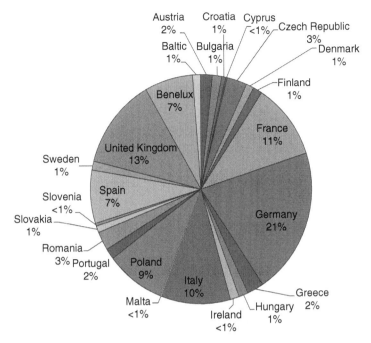

Figure 5.2 National EU emissions 2012
Source: EEA (2014) 'Annual European Union greenhouse gas inventory 1990–2012 and inventory report 2014' Technical Support No 9/14. Available at: http://www.eea.europa.eu/publications/european-union-greenhouse-gas-inventory-2014 Accessed: 28/06/2014.

fairness of proposed commitments, 'possibly against objective indicators' (EU, 2013a, p. 2).

Relative responsibility – The Brazilian proposal

This approach, first proposed by Brazil, in July 1997, as part of the debate on the nature of QELROs to be inscribed in the Kyoto Protocol, has been intermittently developed and discussed. It was reintroduced at the SBSTA in modified form, with the support of the G77, in 2013, but in 1997 it was rejected by the developed countries (Morales, 2013). The Brazilian proposal squarely addresses the issue of fairness with respect to historical responsibility for climate change and provides a definition of 'differentiated responsibilities' and 'equity' under the Convention. In 1997 it called for Annex I countries to reduce their GHG emissions by 30 per cent below a 1990 baseline by 2020. Its key feature was the suggestion that the relative emissions reductions of developed countries should be determined according to their historic contributions to the stock of GHGs present in the atmosphere, associated with the current rise in global mean temperature above pre-industrial levels (Figure 5.3).

Estimating the various national shares involved complex and inevitably controversial methodological issues. The original indicative method led to numbers that assigned the greatest responsibility to those countries that were the first to industrialise. The 'indicative target for the United Kingdom was 66 per cent below 1990 levels by 2010, while the targets for the United States and Japan were about 23 and 8 per cent, respectively' (La Rovere et al., 2002, p. 159). Another part of the proposal involved penalties to be imposed upon Annex I countries that failed to meet their reduction obligations, a variant of the 'polluter pays' principle to be found in EU environmental legislation. The income so generated would be transferred to non-Annex I countries to assist with 'clean development' projects. Elements of this idea subsequently metamorphosed into the CDM. In the original Brazilian proposal there was another calculation of the relative contributions of developing countries, but in this instance those with the largest emissions, notably China, would receive a proportionately large share of the available funds (ibid., p. 160).

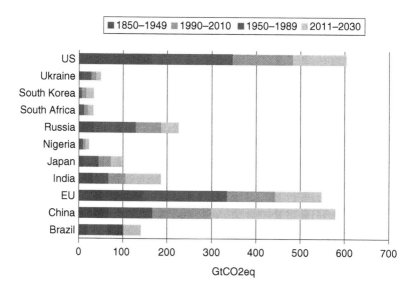

Figure 5.3 Historic cumulative emissions 1850–2030
Source: OECD StatExtracts (2014) Greenhouse Gas Emissions. Available at: http://stats.oecd.org/Index.aspx?DataSetCode=AIR_GHG Accessed: 11/07/2014.

One immediate response to these proposals was that the current governments and peoples of Annex I should not be expected to bear responsibility for actions that took place as much as a century or more ago, sometimes within political boundaries that no longer exist. A reasonable response to this objection would be to point out that, in general, current generations have benefited from accumulated emissions, the by-product of historically unprecedented wealth creation, which in turn laid the foundation for current high standards of living. There is also the claim that the actions of those who built the fossil fuel burning economies of the industrial revolution were unintentional, in that they could not have known about the climatic consequences. However, as Henry Shue (1999, p. 535) points out, while it is not fair to 'punish someone for producing effects that could not have been avoided ... it is common to hold people responsible for effects that were unforeseen and unavoidable'.

The most telling arguments against the Brazilian proposal, in its various forms, have been methodological. It is an extraordinarily difficult task to establish the necessary parameters and, in particular, the relationship between current emissions, atmospheric concentrations of GHGs and temperature change (La Rovere et al., 2002, pp. 161–5). In 2013 the Brazilian proposal was modified to include a reference to IPCC, which would be tasked to establish authoritative and scientifically based 'reference methodology on historical responsibility' to guide domestic consultations for the 2015 agreement (ENB, 2013, p. 26). In rejecting the proposal, the United States pointed out that temperature was a 'lagging indicator' and would 'provide some countries with cover to act in a manner that is much less ambitious than their current capabilities' (Morales, 2013). Switzerland and the EU, while not denying the significance of historical concentrations of GHGs, emphasised the need for a broader range of indicators, including present and future emissions and different capabilities (ENB, 2013, p. 26). This discussion illustrates the near impossibility of achieving a scientifically based consensus free of political taint, compounded by the time pressure of meeting the 2015 deadline. Perhaps a more generally acceptable relative responsibility approach would factor in the growth trajectories of major developing economies in terms of avoiding the 2 °C threshold. It would also need to reference the other key equity principle of current relative capabilities and the ability to pay for the necessary emissions reductions.

Per capita entitlement to a global carbon budget

This approach rests upon the idea that the global carbon budget, the emissions that can be safely made in the future, should be equitably shared. The London-based Global Commons Institute (GCI), a small but highly influential group led by Aubrey Meyer, has been campaigning on issues of ecological debt and climate justice since the early days of the climate regime in 1990. In 1996 the GCI first proposed the idea of 'contraction and convergence', as a means to 'achieve climate justice without vengeance'. Versions of the contraction and convergence idea have been endorsed by governments, including those of the United Kingdom and China, and by the UNFCCC secretariat (GCI, 2011). There are some significant similarities with another budget-based approach, which emphasises equitable access to the remaining 'carbon space'. Carbon space analysis has been developed by Indian scientists and supported by the Indian government, although it draws upon ideas that have widespread currency among research institutes in Europe and elsewhere (Tata Institute of Social Sciences, 2010).

Both 'contraction and convergence' and 'carbon space' models share a basic assumption that there is a limited amount of carbon that can still be added to the existing atmospheric stock. The outer limit is defined by the GCI as the 'full term global greenhouse emissions contraction event that is inevitably required for UNFCCC compliance' or, in plainer language, 'how much carbon consumption is still safe globally' (GCI, 2011, p. 3). This is also the remaining 'carbon space', part of which is, in turn, the 'development space' potentially available to poor countries while still avoiding the 2 °C threshold of dangerous climate change (Figure 5.4). The IPCC 5th Assessment report calculates that, in order to have a more than 66 per cent probability of avoiding a 2 °C rise in mean temperatures, the cumulative amount of carbon in the atmosphere will have to stay below one trillion tonnes – 1000 gigatonnes of carbon (GtC). Lower probabilities allow for greater quantities of carbon, >33 per cent = 1570 GtC and >50 per cent 1210 GtC (IPCC, 2013, p. 27). In mid-2014 the estimated cumulative total was around 580 billion tonnes which, under 'business as usual' assumptions, would mean that the trillionth tonne would be emitted some time in 2040 (www.trillionthtonne.org). Even the inclusion of the one trillion tonnes estimate in the IPCC's policymaker summary was apparently the subject of a difficult debate during an IPCC meeting at Stockholm in 2013, when India and Brazil argued that developed nations were 'using science to legitimize a back door cap on their emissions' (Pearce, 2013, p. 1).

Both approaches rely upon the calculation of *per capita* emissions (Figure 5.5). Under contraction and convergence they are used to estimate equal emission rights for each individual human being at the point of convergence. But the per capita emissions figures are used to

100 *Climate Change in World Politics*

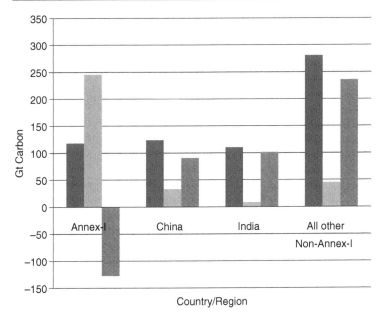

Figure 5.4 Remaining carbon space (entitlements)
Source: 'Global Carbon Budgets and Burden Sharing in Mitigation Actions – Summary for Policy Makers', *Conference on Global Carbon Budgets and Equity in Climate Change 28–29 June 2010, Ministry of Environment and Forests, Government of India*, p. 10.

calculate national responsibilities or entitlements. This is not necessarily the case for the operation of a contract and converge system, but there is recognition that, in practical terms, national and even regional groups would be required to negotiate about aggregate emissions. In allocating the 'carbon space', fair shares are explicitly based on the size of national populations. The 'minimal notion of equity' is 'equal division of the available carbon space among all nations based on their respective populations' (Kanitkar et al., 2010, p. 40). The point at which they were counted (2009) inevitably raises political issues with varying estimates of future population growth and stabilisation.

The models do not focus on individual human rights in ways that might be favoured by cosmopolitan thinkers. Building truly

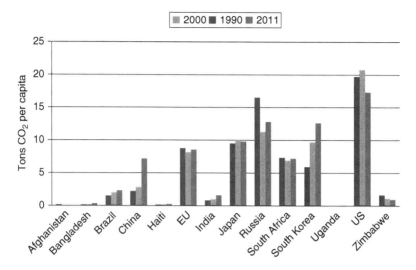

Figure 5.5 Distribution of per capita emissions by country
Source: EDGAR (Emissions Database for Global Atmospheric Research) (2012). 'CO2 time series 1990–2011 per capita for world countries' http://www.themasites.pbl.nl Accessed 11/07/2014.

cosmopolitan carbon budget approaches would pose enormous technical and methodological problems, but these might not be insurmountable if personal carbon allowance systems in particular countries were linked to a global system. The main obstacle is undoubtedly political. A cosmopolitan approach would shine a spotlight on the extreme levels of economic inequality (correlating closely with carbon emissions) within societies. It would accentuate the disparity between the very high personal emissions of, say, Indian millionaires and the minimal emissions of the poverty-stricken masses with no access to electricity or motor transport, which are hidden by nationally aggregated statistics. The same point can, of course, be made for developed societies and for the way in which serious consideration of personal carbon accounting and trading has receded (Seyfang et al., 2009). Discussion of individual entitlements also opens up ethical questions over the priority that should be accorded to 'basic' rather than 'luxury' emissions for the poorest. Ensuring equity in the treatment of the poorest and most vulnerable individuals aligns with a cosmopolitan human rights agenda that seeks to protect the oppressed (Shue, 1995; Elliott, 2006).

Attempts to determine shares of a limited carbon budget involve a zero-sum problem. It is here that the two approaches are essentially different. 'Contract and converge' avoids accounting for the past and proposes a convergence towards equal shares, in actual emissions and entitlements, starting from present levels (flow rather than stock approach). For the developed countries there is, of course, a huge disparity between current emissions and their fair entitlement. The issue of historical responsibilities would be resolved through what is described as the 'main equity lever', negotiating 'a rate of convergence that is significantly accelerated relative to the rate of contraction' (GCI, 2011, p. 3). While accepting the principle of 'contraction and convergence', the Chinese government has argued that the rights of developing countries are being continually infringed by Annex I economies and that there should, therefore, be immediate convergence, with the gap between actual and entitlement emissions being balanced by the trading of quotas (ibid., p. 5).

For 'carbon space' advocates, the all-important issue is its 'over-occupation' by the developed nations. Current negotiating approaches based on achieving reductions in emissions flows at 2020 or 2050 are unacceptable, because:

> While the gross inequalities in per capita emission flows certainly indicate inequitable access to the global atmospheric commons, it does not provide any justification for the continued emissions by the developing countries in order to realize their development goals.
> (Kanitkar et al., 2010, p. 40)

Analysis of cumulative carbon stocks over a long historic period, either from 1850 or from 1970 to 2050, defines the carbon space. Using IPCC scenarios it transpires that Annex I countries have already used their fair share and are in substantial deficit, while developing countries should be allowed to utilise the major part of the remaining carbon space. On a 1970 basis, this deficit is calculated as a 2009–50 emissions entitlement for Annex I countries of minus 100.38 GtC. The remaining entitlements of China and India are 79.97 and 99.17 GtC respectively, with the rest of the world allowed 222.24 GtC (Kanitkar et al., 2010, p. 50). The policy implications for Annex I countries are so severe that carbon space analysis is more a means of driving home a political point about past inequities and future barriers to sustainable development than the basis for a practical negotiating programme.

Duties of mutual aid and responsibility for loss and damage

In signing the Convention, developed world parties recognised that they had a duty under Article 4.4 to provide assistance to developing countries particularly vulnerable to climate change to help meet the costs of adaptation. According to Adger et al. (2006, p. 16) this potentially met one distributional requirement of climate justice. Another procedural aspect was envisaged in Article 4.3, which required developed countries (in Annex II) to provide 'new and additional financial resources to meet the agreed full costs incurred by developing country parties in complying with their obligations'. As the regime developed, and certainly after the 4th Assessment Report of the IPCC in 2007, it became clear that mitigation efforts alone would be insufficient to avoid the damaging effects of climate change and there would have to be a new focus on adaptation. This would seek to respond to the evident injustice of climate change visited upon those developing societies that had virtually no responsibility for causing the problem, but which were likely to encounter its worst effects. Most of these vulnerable, frequently tropical or sub-tropical, countries were also the least well-equipped to adapt, in terms of finance and technical capabilities. This much, at least, was relatively uncontroversial. Starting with the Bali Plan of Action in 2007 and extending through the decisions at Cancun in 2010 and beyond, a governance structure to enable adaptation was slowly constructed. It contained both aid and technical elements, although funding was tardy, plus an attempt at ensuring procedural justice with the establishment of an LDC expert group and provisions on public participation in the drawing up of National Adaptation Plans. This involved some slight movement towards calls for the recognition of issues of domestic climate justice and the need to provide space for societies to develop their own adaptation strategies (Thomas and Twyman, 2004). In parallel there was also an ongoing negotiation on deforestation and REDD+. It too raised ethical questions of responsibility and equity but analysis of the evolution of proposals on 'avoided deforestation' suggests that neoliberal market-based principles have tended to predominate (Okereke and Dooley, 2010).

The provision of 'the means to address loss and damage associated with climate change impacts in developing countries that are particularly vulnerable to the adverse effects of climate change' (UNFCCC, 2009) was also placed on the agenda in the Bali Plan of Action. It was to raise some difficult and pointed issues of responsibility and justice. The essential idea was that highly vulnerable societies subject to imminent climate disasters and 'slow onset events', and lacking both capability and developed insurance arrangements, should receive international support. There were tortuous discussions and, by the Durban COP of 2011, three areas had emerged for consideration: risk assessment, approaches to address loss and damage and the role of the UNFCCC in enhancing implementation (Warner and Zakieldeen, 2011). Lying behind them was a North–South disagreement on the question of responsibility and compensation. After Bali, developed countries spotted that AOSIS proposals on the issue included a 'rehabilitation and compensation' component, which implied that they might be held responsible for climate change 'loss and damage'. To avoid this they attempted to avoid discussion of anything other than risk management, insurance and capability building, or to include the loss and damage issue in the general adaptation discussion (ibid. p. 4). Arguments on this question of the separability of loss and damage from adaptation were still going on in Warsaw at the end of 2013, where 'developed countries stated that loss and damage is part of the mitigation and adaptation continuum, whereas developing countries identified loss and damage as a separate issue, distinct from adaptation' (ENB, 2013, p. 18). This apparently trivial, but significant, disagreement was the occasion of a walkout from the COP by over 130 developing countries, with the Like-Minded Group in the lead (Byravan and Rajan, 2013).

At stake is the key issue of responsibility for climate change and liability for compensation to the victims, which could be very substantial. According to Munich Re, direct economic losses (totalling $100 billion per annum since 2000) were most severe, in relation to national income, in low-income countries. Also, if low-lying areas and small island states are inundated, then the acceptance of responsibility could involve migration and re-settlement of populations. It is, of course, difficult to apply existing international legal principles of 'no harm' or the 'polluter pays' to climate change events where there is no clear agency, but rather the accumulation of the diffuse effects of a myriad of human activities over a long period. Nevertheless, developed countries are clearly the major culprits and links have been made to arguments about their occupation of the 'carbon space' (see Figure 5.4) where 'paying for loss and damage would maintain equity with regard to the current stock of carbon dioxide' (Byravan and Rajan, 2013). In an interesting reversal of the usual flow of culturally specific ideas, Bangladesh-based advocates of a compensation regime invoke, not only 'Northern precedents for international compensation arrangements', but also European and North American legal principles of tort and liability (Khan et al., 2013). On the basis of established international compensation schemes, the most viable proposal would be some form of international fund for the compensation of climate victims.

While there has been general acceptance by most rich, developed countries for an obligation to provide aid funding for adaptation and for some aspects of risk and disaster management, this is not regarded as reparation for past damage. US lead negotiator Tod Stern accepted that there is 'no question that we need to provide assistance to many countries that are working to build low-carbon economies and many countries seeking to build resilience and to adapt to climate impacts'. However, 'we need to elevate practical problem solving above rhetoric and ideology. Lectures about compensation, reparations and the like will produce nothing but antipathy among developed country policy makers and their publics' (Stern, 2013, p. 7). From a Bangladeshi perspective, poor people already carry a heavy burden:

> Without their being effectively compensated for additional loss and damage caused by the wealthy societies in the North there is no justice – and no new agreement on climate change!
> (Khan et al., 2013, p. 5)

Conclusions

Contrary to the advice of commentators, who reflected on the difficulties of attempting re-distribution under the common heritage provisions of the Law of the Sea Convention or in the New International Economic Order (NIEO) debates, considerations of justice were part of the climate regime from its inception. There was always an underlying understanding that a 'sustainable development' bargain would have to be struck, in which the environmental requirements of the North could only be met by accommodating the development needs of the South. In one sense the 'macro rationale' for the climate regime must be deeply cosmopolitan, with a conception of one earth with one climate inhabited by rights-bearing human individuals obligated to each other.

The actual implementation of the climate regime bears the hallmarks of a communitarian and pluralist approach, which may not have had entirely negative implications, in the sense that pluralism places great stress on appreciation of, and respect for, international diversity and difference. The Kyoto Protocol required mitigation action by developed countries that accepted the obligation of making the first move. However, since 2001, many Umbrella Group members have abandoned this position and openly oppose CBDR-RC, while claiming 'fair' treatment with respect to the major developing economies of the South. The recent arguments over 'loss and damage' are part of long-standing disagreement over responsibility for the harm of climate change, although the very fact that these issues are discussed at all may represent some form of, agonisingly slow, progress.

It is easy to be cynical about the versions of distributive justice put forward, and frequently rejected, in the negotiations. In most cases they align quite closely with the national economic interests discussed in Chapter 4. Thus, for example, India would receive substantial advantage from the 'carbon space' approach it has sponsored, on the basis of its low historic emissions but large and expanding population. The EU, and even China, see advantages in 'contract and converge', but with different positions on the rate of convergence. No one could blame AOSIS for demanding restitution for loss and damage. The denial of equity by the US hardly merits comment. Realism, which emerged as a distinct position in international thought as a reaction to the 'idealistic' schemes of the League of Nations and the Kellogg-Briand Pact, would find the fate of contemporary schemes for a fairer international climate order distinctly unsurprising.

However, this is not the entire story. The EU has taken up positions on climate justice, which are not necessarily in direct alignment with the national energy and economic interests of its member states, and has displayed a willingness to seek solutions within the CBDR-RC formula and to provide substantial climate-related adaptation funding (as have other Annex I countries, while denying that this derives from strict responsibility for accumulated climate harm). Participants in the Cartagena Dialogue and the members of AILAC are similarly willing to cross the ideological lines of entrenched positions on climate justice. Ultimately, however, it is the nexus between national interests and demands for justice that is the key determinant of future progress. There will not be an effective climate agreement without recognition of inequalities and a means of dividing burdens and providing funding that are accepted as sufficiently fair by all participants.

There is another political dimension to the justice debates. As ideas of justice are deeply held, providing legitimacy and inspiration far beyond what can be achieved by a simple appeal to self-interest, they are important in mobilising coalitions, enhancing and even destroying the credibility of Parties and, therefore, a powerful political tool in their own right. It is to this reputational aspect of international climate politics that the next chapter will be devoted.

6
Recognition and Prestige

The very earliest writings on international relations confirm the significance of the pursuit of honour and prestige alongside more 'base' concerns with relative power and wealth. Thucydides's description of the Peloponnesian war accounts for the fate of the Melians in their unequal struggle with the Athenians. Simple survival should have counselled surrender, yet honour dictated what turned out to be a suicidal course of action. This theme is taken up in classical realism. In Martin Wight's (1978, p. 97) discussion of power politics 'honour is the halo around interests, prestige is the halo around power'. Hans Morgenthau (1967, p. 69), doyen of realist theorists, identified the contest for prestige as one of three 'basic manifestations' of the struggle for power in international relations and outlined the prestige policies that statesmen may pursue. The other two are protection of the status quo or imperialism – where pursuit of prestige represents one of the instrumentalities through which they may be achieved.

For realists prestige is a 'positional good' and status competition has a zero-sum character. In a very different formulation, Richard Ned Le Bow (2008) in his *A Cultural Theory of International Relations* treats what he calls 'standing' as an expression of a universal human motivation to achieve self-esteem, which is measured and validated through social culture. This is both an end in itself and a means to various political ends.

What had been the honour of sovereign princes became, in an era of nationalism, a matter of sacred national duty. The events preceding the First World War spring to mind – the French refusal to make 'rational' adjustments to the loss of Alsace Lorraine and the bombast of Willhelmine Germany – alongside more recent examples such as the 1982 Falklands/Malvinas conflict. The latter provides a good example of non-material interests. There were few grounds for resisting the Argentine seizure of the islands based on any concept of rational or material interest. Indeed, the UK government had been seeking for some time to find a means to transfer sovereignty to Argentina that would be acceptable to the tiny population of this British dependency. It has been revealed that Mrs Thatcher's economic adviser, Sir Alan Walters made what seemed to him the perfectly reasonable suggestion that the Falkland islanders be offered $100,000 per family plus a lifetime guarantee of settlement and full citizenship in Britain, Australia or New Zealand, if only they would accept Argentine sovereignty over the islands (*Guardian*, 2013, p. 1). However, the affront to British prestige entailed by a violation of its sovereignty was such as to transform the situation. It was evident that, had the Thatcher government not been able to launch the task force to retake the islands, it could not have survived. Expenditure cuts and reductions in the navy's surface fleet, which had been at the centre of recent domestic political argument, were forgotten. They no longer mattered, for how could one calculate the cost of 'national honour'? The successful conduct of the war brought, of course, electoral dividends to the Thatcher government.

Such events are at the extreme end of the scale, but everyday diplomacy has always been marked by status concerns and competition. They often lead to arcane questions of protocol, precedence and ingenious forms of one-upmanship. Sensitivity to questions of national recognition by other states is, thus, ever present in diplomacy, down to micro levels of interaction. The climate regime is no exception to this rule. The appointment of a state as host and president of a COP is a matter of national prestige. This was evident, for example, in Japan's enthusiasm to host the Kyoto COP and in the amount that countries will invest in holding such large and high-profile international events. In 2015 'Team France' for COP 21 involved the full-time efforts of some 60 civil servants led by the Foreign Minister for what was clearly a high national priority (French Government, 2015). There can also be a certain proprietorial pride in the naming of agreements: the Bali Programme of Action; the Copenhagen Accord; Durban Platform or Doha Gateway, which may have some political effect in terms of the gains that may be registered from a successful negotiating outcome. Certainly the role of a national presidency of a COP is regarded as an important opportunity to demonstrate concern, diplomatic competence and leadership. In practical terms it is far from being a purely honorific position. As outlined in Chapter 3, the host nation's presidency has an important role in orchestrating negotiations, convening high-level informal meetings and generally acting to ensure the required positive outcome from the Conference. The failure of such a well-publicised international negotiation to produce the appropriate outcome documents and to demonstrate progress would reflect poorly on the international standing of the host.

Climate change and the politics of prestige

The relevance of climate diplomacy to international status competition may not be immediately evident. Most day-to-day international environmental negotiation – on hazardous waste, biodiversity or a multitude of other subjects has a functional character. The bulk of climate diplomacy in the subsidiary bodies and in the various working groups and committees can be described in this way. The agendas are often highly technical and the participants are usually specialised, national civil servants from environment and energy ministries. However, the 'high level segment' that is scheduled for the end of COPs provides a public stage attracting significant media attention and, usually exaggerated, declarations of 'breakthrough' and 'deadlock'. The participants are normally the environment ministers of the Parties but, on occasion, heads of government and the UN Secretary General will attend to emphasise the significance of the negotiation and to be seen to be taking decisive action. The signature of the Convention, at the 1992 Earth Summit, was such an occasion. Among the 10,000 participants at the 1997 COP at Kyoto were the prime minister of Japan, the presidents of Costa Rica and Nauru and US Vice President Al Gore, along with 125 environment ministers (ENB, 1997, p. 1).

The 2009 Copenhagen COP 15 was promoted as the moment at which a new comprehensive agreement, supplanting Kyoto, would be established. A concerted campaign was mounted to raise the status of the COP, in which first one then another political leader indicated their intention to attend, turning it into something tantamount to a world summit meeting. The recently elected US President Obama, European presidents and prime ministers and the leaders of the BASIC countries were present. In all, 119 heads of government attended (Kanie, 2011, p. 117). Having reached a deadlock on the final day of the conference, a 'Friends of the Chair Group' of 25, of which the majority were heads of government, was assembled. They laboured to produce the necessary agreement – the Copenhagen Accord – while regular negotiations were 'virtually suspended' (Dimitrov, 2010, p. 809). The final obstacle to the

Accord was dealt with in a closed session between the leaders of the United States, China, India and Brazil (ibid., p. 810). If testament were needed as to the elevated political status of climate issues, Copenhagen provided it. In fact this had already been indicated by the inclusion of climate change as a key agenda item at G8 summit meetings, most notably at the Gleneagles meeting of July 2005, and also by the convening by UN Secretary General Ban Ki Moon of a climate summit meeting in New York, in September 2009, as a precursor to the Copenhagen negotiations.

After 2009, and in the context of the downturn in the world economy and political turmoil in the Middle East, the international salience of climate issues tended to recede, but there were strenuous efforts by European leaders and by the UN Secretary General to keep climate on the agenda of political leaders. Once again Ban Ki Moon convened a New York climate summit in September 2014, preceded by a large public demonstration in which he personally took part. This time the summit meeting, designed to advance negotiations through high-level political involvement, left a substantial period of 14 months before the scheduled signing of a new climate agreement in Paris.

Shifts in the framing of climate issues, discussed in Chapter 2, are relevant to the way in which climate change has achieved greater political salience than other environmental issues. The 'securitising' moves of the British and German governments in 2007 and 2011 at the UN Security Council can be read as conscious attempts to promote climate as 'high politics'. In realist thinking, the 'high politics' of statecraft has been associated with core issues of national security and international power competition, which are the proper subject of international political analysis. 'Low politics', relating to functional cooperation and the performance of necessary but mundane international tasks, such as the regulation of telecommunications, infectious diseases and transboundary pollution, were of little interest to statespersons. In some realist-inspired analysis, 'climate change has become a matter of national security'. It was 'no longer a simple environmental and sustainable development issue', but 'a serious issue with potential negative consequences for both the United States and its allies':

> It could change territorial boundaries, impact patterns of migration, change the livelihoods of millions, if not billions of people across the globe, and simply be the main game-changer of the 21st Century to the international order.
>
> (Motaal, 2010, p. 105)

The close relationship to energy policy must also be significant here. Conflicts over energy resources were a *leitmotiv* of twentieth-century power politics. Indeed, they are likely to become even more significant in the context of dwindling reserves and the complicated intersection of climatic change and shortages of other essentials, such as clean air and drinking water. What might be decided, or avoided, in climate negotiations could have serious implications for the economic fortunes, and hence the relative power, of the major players in international politics. However, although climate change is inevitably a factor in the long-range strategic planning of governments, this does not necessarily mean that it can be treated as a 'high politics' issue in its own right, alongside more orthodox conceptions of national security. Small island states where there will be immediate questions of national survival are an exception, but elsewhere climate change will only be associated with core 'high politics' issues of national power and security.

The ascription of status

The implications of the politics of status and prestige are likely to go well beyond the specifics of any negotiation. They could involve establishing the legitimacy and identity of states, achieving recognition and asserting leadership. Leadership both relies upon and confers status. Its exercise often depends, not only on the possession of economic or political resources, but also on appeals to climate justice. This helps to explain the disproportionate influence of AOSIS in setting the negotiating agenda. There is also the important question of how the climate regime fits with, and is perhaps moulded by, the broader status competition between major powers in the international system.

From a sociological perspective on international relations, there can be two related sources of status ascription. The first is self-referential, relating to national traditions, memories of historic grandeur and aspirations to future great power status. There may also be a potent desire to erase past humiliations and to demonstrate independence and significance with respect to past colonial overlords. The second is 'community attribution' (Volgy et al., 2011, pp. 6–8). This will frequently be indicated by membership of exclusive international clubs. The obvious example is a permanent seat at the UN Security Council, but there is also inclusion in the list of G8 and G20 members. Community status attribution can be important at the level of regional or other coalitions, including the G77, the BASICs and the specific climate-related alliances outlined in Chapter 4.

The significance of recognition, reputation and the avoidance of insult may be easily grasped in the everyday experience of individuals in social life. What may seem implausible is the transfer of such individual meanings and motivations to the collective or state level. There is a telling discussion in Hollywood's celebrated anti-war film of 1930 *All Quiet on the Western Front*. As a group of weary German infantrymen rest in a field, they discuss the causes of the conflict in which they have become so intimately involved. One of them refers to national honour. His comrade points out the total absurdity of it all, '... as if a hill in France could insult a mountain in Germany!' In this sense, it is inaccurate to endow the state with feelings that can only be possessed by a human individual. However, this is to neglect 'the "affective" and even the "identity" value that an abstract institutional identity – even if it is highly "fictitious" – can possess for officials of such an institution' (Lindemann, 2010, p. 18).

For much of recorded international history there was a more direct, personal connection, as diplomatic activity was entangled with the hierarchical sensitivities of sovereigns and the aristocratic castes by which they were surrounded. Today, it is evident that wider publics also internalise national identities, frequently associated with sovereignty and the legitimacy of particular regimes. Hence, governments will seek to improve their standing with their own public in ways which, although not commensurate with victory on the battlefield, serve to embellish national prestige – for example the staging of high-profile events such as the Olympic Games, or even the hosting of G8 summits. The annual climate COPs are, as noted in Chapter 3, a site of significant political activity by environmental NGOs. Much of this is clearly targeted at media coverage that will emphasise the overall significance of the issue, but also 'name and shame' reluctant governments. Public attention across the political systems of the major state participants in international climate politics varies, and is subject to the fluctuations of the 'issue attention cycle'. In North West Europe there have been relatively high levels of public interest in environmental issues. Yet in much of the Anglo-Saxon world there has been an orchestrated campaign or 'social counter movement' to deny the science, and hence the significance of climate action, leading to an outright hostility in the US Congress and elsewhere (Jacques, 2012). This has been a decisive factor in US climate policy since before the signature of the Kyoto Protocol, and limits or even nullifies the domestic political benefits that might flow from high-profile international action on climate by a US president. For some vulnerable Parties, climate can be represented as *the* issue of concern for governments, while, for fossil fuel exporters, domestic kudos may rest upon the extent to which governments can stymie international action. There are marked differences between the advocacy of Northern and Southern NGOs, with the latter inevitably prioritising development and forestry issues (Doherty and Doyle, 2013). NGO advocacy can tap into the reputational concerns of governments. Much of the latter's status-seeking behaviour is for domestic consumption and serves to legitimise regimes and build domestic support. In negotiations, officials will know that their own careers may be harmed by poor performance, defined in terms of loss of reputation, and they will very likely have internalised ideas about the proper position and respect to be granted to representatives of their government. The same will apply, in a much more public and high-profile way, to ministers.

The politics of prestige and recognition is not merely a matter of domestic political advantage. It also plays a deeper role in identity formation. This point has been emphasised by constructivist scholarship in IR (Wendt, 1994, 1999; Hopf, 1998; Hopf, 1998). From this perspective, states are involved in a continuous process of identity construction in relation to other members of the international system. This intersubjective structure of meaning provides the essential context within which interests are conceived. It contrasts with the fixed notions of national interest employed in more conventional views of state motivation and, indeed, with the idea that prestige assumes some form of quantity that can be amassed. The formation and maintenance of identity may be of particular importance for new states or those that have emerged from a long period of dependence and subjugation. The assertion of major power status by China and India over recent decades is apparent and, it will be argued, constitutes an important component of national positions on climate change.

An interesting, but rather different, instance is provided by the EU and its aspiration to exercise leadership in the climate change regime. Here, it is the national traditions of the various member states that have to be combined, even over-ridden, in the creation of a distinctive international actor, visibly separate from its component parts. While the EU has been a very significant single actor in international trade politics, on the basis of the Treaty requirement to have a common external tariff, it has been less evident and effective in other aspects of its external relations, notably the Common Foreign and Security Policy. External environmental policy in general and climate policy in particular, has been a notable exception (Bretherton and Vogler, 2006, pp. 89–110). Thus, the record of EU climate leadership, its spokespersons have self-consciously claimed, is also a story of identity formation for the EU. As it is logically impossible to identify a 'national interest' for the EU, this case also highlights the way in which identity formation and interest are intertwined in the formulation of foreign policy.

Recognition and the assertion of sovereignty

Since the break-up of colonial empires at the end of the Second World War, the number of state units in the international system has virtually quadrupled. The increase was particularly marked during the 1960s and 1970s, leaving in its wake a large number of underdeveloped and weak states in a position of substantial disadvantage within an international economic system that cast them in a dependent role as raw material exporters. The governments of these states were frequently wary of their former colonial masters and of the encroachments of transnational capitalism. Above all, they were concerned to preserve a new and fragile sovereignty. The international campaign for a New International Economic Order during the 1970s was a collective manifestation of the demands of the developing world, expressed through the United Nations. It influenced the form of campaigns to establish 'common heritage' principles in the Law of the Sea and was evident in environmental politics from the 1972 Stockholm Conference through to Rio, 20 years later. Economic sovereignty over natural resources, in particular tropical forests, was a central demand. Stephen Krasner (1985) interpreted this 'structural conflict' as a rejection of global liberalism, driven by the need to preserve sovereign independence: 'The meta political goals of third world states, and many of their relational goals as well, can be understood by reference to the minimalist objective of preserving political integrity' (Krasner, 1985, p. 28).

Much has changed since the 1980s, but some elements clearly persist in the understandable sensitivity of many developing states to potential infringements of their sovereignty. This is not to suggest that the concerns of highly vulnerable countries, and their requirements for adaptation finance, are anything but genuine. However, it should alert us to a range of essentially political motives and assertions of independence. These are evident in discussions of monitoring, reporting and verification (MRV). Umbrella Group members tend to regard intrusive MRV provisions for nationally appropriate mitigation actions (NAMAs) as an important way to prevent 'free-riding', to ensure fair competition with economic competitors and to achieve transparency so as to police the use of adaptation funding.[1] Non-Annex I countries and funding recipients, however, often regard MRV as an affront to, and even a potential transgression of, their sovereignty. Even China, a global power, was sensitive to sovereignty issues and mindful of past subjection and humiliations, although its attitude to MRV mellowed somewhat once it had been invited to participate in reviews of Annex I countries. The important point is that there was no problem as long as China was 'treated respectfully and as an equal partner' (Zhang, 2003, p. 77). Demands for strict monitoring standards are especially galling when '… the contribution of developing countries to mitigation efforts is far greater than that by developed countries … the pre-2020 mitigation gap would not even have existed if developed countries had committed to an emission reduction of 40 per cent below their 1990 levels by 2020' (BASIC, 2013, p. 2).[2] Another set of sovereignty-related issues involve suspicion over the extension of market-based approaches, unfair barriers to technological change contained in intrusive intellectual property rights (IPR) rules and worries about the extent to which climate funding will be provided by an unaccountable private sector (LMDC, 2013, p. 3).

The actions of ALBA at the Copenhagen COP present the most striking case of radical dissent in the history of the climate regime. At least three aspects of status politics were present: an appeal to their domestic public and anti-imperialist tradition, an assertion of their sovereignty and collective identity with respect to the United States and other Latin American countries and a gesture of resistance against the imposition of an agreement cobbled together by a big power 'club'. The Bolivarian Alliance comprises Bolivia, Cuba, Ecuador, Nicaragua and Venezuela (along with the Caribbean islands of Antigua and Barbuda, Dominica and St Vincent and the Grenadines). Created by an agreement between Presidents Hugo Chavez of Venezuela and Fidel Castro of Cuba, it seeks to realise Simon Bolivar's dream of a union of Latin American countries capable of fending off the might of their giant neighbour to the North. Specifically, ALBA has attempted to organise an economic alternative to US proposals for a Free Trade Area of the Americas, based on socialist principles and mutual aid. The latter has involved exchanges of oil and medical personnel. ALBA governments have adopted a fiercely anti-imperialist stance. The alliance was hardly designed with environmental policy in mind, rather a climate change stance has been adopted that fits its general ideological orientation. Thus environmental degradation is seen to result from the 'crisis of world capitalism', while the struggle against capitalism and unbridled industrialisation equates to defence of 'mother earth' (Janicke, 2010).

ALBA's moment came at the very end of the Copenhagen COP, when 'friends of the chair' devised a compromise text, acceptable to the US and the BASICs, and when the presidents and prime ministers, having announced their agreement, were already on their way home. In the early hours of Saturday morning, in the final plenary session, ALBA countries objected to the text. As Radoslav Dimitrov (2010, pp. 810–11), a participating delegate, notes, other delegations also objected openly or implicitly, but it was ALBA that seized the international stage. Venezuela called the Accord 'a coup d'état against the Charter of the United Nations' and Cuba termed the aid package 'disgraceful' blackmail. In what Dimitrov (2010, p. 813) describes as 'the most dramatic episode of world politics in action that I have ever witnessed ... around two hundred people gathered at the center, surrounding delegates from Bolivia and Venezuela who were physically pressed against the wall of the podium'. The consequence was worldwide publicity and the failure by the COP to adopt the agreement that was, in consequence, known only as the Copenhagen Accord.[3]

ALBA's position was one of ideological declaration. President Evo Morales of Bolivia, for instance, called for an impossibly ambitious target of a 1 °C temperature increase and repeated Fidel Castro's claim at the 1992 Rio Earth Summit that capitalism was the cause of climate change (*Guardian*, 2009). The process of negotiation was declared to be undemocratic, but 'above all' it failed to recognise the principle of 'sovereign equality between all countries'. 'We, the developing countries, are dignified and sovereign nations and victims of a problem that we did not create' (ALBA, 2009).

Climate policy leadership

The climate regime seems particularly prone to assertions of leadership, even when there may be a distinct lack of followers![4] Brazil, India, Japan and, of course, the EU have all claimed to be climate policy leaders. The United States, despite its rejection of Kyoto, has claimed the mantle of leadership. This was an aspiration of the Obama presidency before and after Copenhagen, reasserted in the administration's 2014 climate policy (Gillis and Fountain, 2014). Commenting on the aftermath of Copenhagen, the president of the European Commission was heard to remark, sardonically, 'too many leaders!'[5] Leadership and its various dimensions, whether structural, entrepreneurial or cognitive, has also been the subject of much academic discussion (Wurzel and Connelly, 2010). The focus has been on its effectiveness in the pursuit of international agreement. Rather less attention has been paid to the reputational importance of leadership and why the title of 'leader' continues to provide such a popular self-ascription.

The EU's aspiration to international climate policy leadership dates back to the early days of the UNFCCC and forms part of a broader approach to environmental questions (Vogler and Stephan, 2007). The Union is neither a proto-federal state nor an over-developed international organisation. Instead, it has attempted to form its own unique identity as an international actor, at once including and distinct from its 28 member states (Bretherton and Vogler, 2006). As its efforts to develop an effective Common Foreign and Security Policy have been constrained by the need for intergovernmental consensus and the essentially civilian character of the Union, environmental and climate policy have assumed major significance. The EU's long-standing trade and development aid relationship with African and other developing states is another significant part of the Union's external role. A typical strategy, in support of 'targets and timetables' under a comprehensive agreement, has been to attempt a mediating role between the North and South, assisted by offers of aid funding. This, too, conforms to an official self-image of a beneficent normative and civilian force in world affairs. This may not be uppermost in the minds of EU citizens beset by the continuing problems of the euro, but it has been recognised by the Nobel Committee's award of Peace Prize in 2012 (a distinction shared with the IPCC and Al Gore, jointly awarded the 2007 Peace Prize for their climate change activities).

Statements by the European Council and EU officials are routinely prefaced by references to the EU's leading role in the formation of the international climate regime and the necessity, in terms of international credibility, to develop and sustain EU internal climate policies, including the flagship ETS. Continued activism on issues such as the inclusion of air and maritime emissions in the ETS, and on the additional taxation of fuels derived from tar sands, signals the seriousness of its intentions but embroiled it in disputes with allies and trading partners. Although transatlantic ties are both extensive and complex, climate change provided an issue on which the Union consciously challenged the United States. When the US government announced its withdrawal from Kyoto in 2001, the European Council decided to proceed with the Protocol. In the face of US obstruction, it proceeded to flesh out an agreement of some complexity and novelty, particularly with regard to organising and ensuring compliance with the CDM. This onerous task was completed during the remainder of 2001, at an additional COP 6 *bis* held at Bonn and at COP 7, which produced the so-called Marrakesh Accords, turning Kyoto into a ratifiable instrument. Various developed countries were already inclined to follow the US lead, and the Union was forced to make substantial concessions to accommodate them. Having developed the Protocol, the task was then to ensure ratification and entry into force. This required not only that 55 per cent of the Parties ratify, but that they must also be responsible for 55 per cent of global emissions. With the leading emitter (the United States) absent, this meant that both Japan and then Russia be persuaded to ratify. In a concerted diplomatic effort, which in the Russian case involved trade-related promises, the Union achieved its aim (Bretherton and Vogler, 2006, p. 109). Having received the necessary ratifications, the Protocol entered into force on 16 February 2005. By any standard, the period between 1997 and 2005 demonstrates consistent political resolve and creative climate policy leadership, only occasionally interrupted by internal dissension.

The broader context included 'normative' conflicts with the US administration, notably over the establishment of the International Criminal Court and the abolition of the death penalty (Manners, 2002). It was a period of European enthusiasm in which the EU was sometimes portrayed as an emergent superpower of a new type, in stark contrast to an assertive yet uncooperative US administration (Reid, 2004; Rifkind, 2004).

The events of 2008–9 put an end to this optimistic phase, but the EU continued to play the role of climate leader, despite the severe setback to its reputation in Copenhagen. Publically side-lined in the negotiation of the Copenhagen Accord, policy towards the United States appears to have been modified. We have a rare insight into the day-to-day diplomatic discussion of symbolic issues in leaked US embassy cables from the early part of 2010, following the Copenhagen COP. For one member state, Spain, a summit meeting during its presidency of the Union in the first half of 2010 was important because of the need to discuss climate change, but '... if the US could commit itself now to a bilateral meeting in Madrid, later this year, then Spain would be satisfied' (US Embassy Cable, 3 February 2010). Later in the same month there was a roundtable meeting of US and EU officials in Brussels, where the disunity of the US and EU in the face of the 'remarkable' and 'closely co-ordinated' action of the BASICs was discussed. The US side stressed that 'EU leaders' one-upmanship model of outdoing each other to push EU-wide policy did not work in dealing with the Obama administration' (US Embassy Cable, 3 February 2010). Climate Action

Commissioner Hedegaard said that 'She hoped that the US noted that the EU was muting its criticism of the US, to be constructive', while the EU Director-General for External Relations asked for US understanding of the political importance of EU–US summits, pointing out that 'symbolism is important to EU institutions' (ibid.).

Climate policy continued to be one of the clearest manifestations of the Union as an effective international actor both externally and internally. Eurobarometer findings suggest that this is in marked contrast to public apathy and hostility towards EU-level action in other policy areas (Adelle and Withana, 2010). The European Council for Foreign Relations publishes an annual scorecard, rating the success or failure of the Union's external actions. Very few, if any, achieve high grades, but climate policy achieved a B+ for the Cancun COP in 2011 and a rare A– for the Durban outcome in 2012, where the Union had '... made diplomatic progress where none seemed likely. Although imperfect, the deal was a significant victory for European diplomacy' (European Council for Foreign Relations, 2012, p. 122).

The EU is the most prominent example of a polity with a demonstrable stake in acquiring status through climate activism. Japan in the 1990s also claimed leadership and enjoyed the symbolic benefits of hosting the negotiations at Kyoto. As with the EU, claims to climate leadership were regularly repeated in official statements, and there was a high level of domestic public support for this activity (Kanie, 2011, pp. 120–5). Climate activism can be interpreted as a means by which Japan could assert major power status consistent with its technological prowess and position among the largest of the global economies. This was already acknowledged in Japan's membership of the G7 group of leading 'industrialised economies', but, as a consequence of the post-World War II settlement, and under Article 9 of its constitution, Japan was debarred from the normal military power projection and nuclear weapons attributes of great power status. The idea, pursued by a succession of prime ministers from 1987, was that there might, under changed circumstances, be a way of asserting Japan's importance and satisfying national aspirations through leadership in the civilian dimensions of international politics 'contributing to international peace and prosperity' (ibid., p. 121).

There was a strong parallel with notions of the EU as a mainly civilian power with very limited access to 'power projection'. However, the policy of climate activism did not extend into the post-2012 discussions, when Japan, unlike the EU, declined to participate in a second Kyoto commitment period. Economic considerations appear to have outweighed the concerns of foreign and environment ministries. While ratifying the Kyoto Protocol, Japan declined to join the EU in leading its implementation. In part this appears to reflect its awareness of US hostility to the Protocol and the centrality of the US alliance to Japanese security in an era of rising tension in East Asia. Also, further reductions in its CO_2 emissions would be disproportionately expensive because of its existing high levels of energy efficiency. A combination of bureaucratic conflicts, possible economic damage and substantial pressure from the US appear to have made 'the pursuit of status in the climate change issue area a lose-lose proposition' (Kanie, 2011, p. 29). Sponsoring the 2010 Nagoya Protocol to the Convention on Biodiversity, and subsequently the Minimata Convention on Mercury, signed at Kumamoto, Japan may have provided a rather limited, yet uncontroversial, alternative to Kyoto.[6]

Community ascription: The great power 'club'

How has the development of the climate regime interacted with overall international status competition? There are several possibilities. It may be that the adoption of a leading role in the climate regime confers standing, or that the existing status hierarchy, demarcated by admission to various exclusive international clubs, has been deployed to influence behaviour within the climate regime.

Much of the writing on international status concentrates on the question of those states that may be regarded as great powers. The data on this appears remarkably consistent from 1945 to 2005. There have been and remain seven great powers: the USA, France, the UK, USSR/Russia, Germany, China and Japan (after a post-World War II interval). In the Correlates of War Project data that is usually employed, status attribution is based on expert views and there is no attempt to rank great powers (Volgy et al., 2011, p. 5). Tom Volgy and his collaborators have attempted to update and develop empirical measures of great power status, which is defined in terms of 'unusual material capabilities, the willingness to pursue a wide range of foreign policy interests across a large geographical area and recognition by other states that they are major powers' (ibid., p. 12). On this basis, they are able to identify status consistent powers, the United States being the primary example, and to trace inconsistencies between status indicators that define 'underachieving' and 'overachieving powers'.

Japan represents an 'underachiever', where the status accorded by the international community is less than what might be expected on the basis of material capabilities. Japan's attempt to emphasise its role in environmental and climate diplomacy can be interpreted as a response to underachievement. The EU is an actor in climate politics, but does not figure in these state-based calculations of international standing. The notable absentees from this list are the BASIC countries other than China, frequently regarded as rising or emergent powers, crucial to the future of the climate regime. In many ways the status competition tracks the structural changes in the global economy that will be outlined in Chapter 7, but the fit is hardly exact. The great power 'pecking order' has historically been associated with military strength, but with the demise of the USSR there are no great power comparators to the United States and the origins of the climate regime coincide with the US 'unipolar moment'. A long-term badge of status remains permanent membership of the UN Security Council, which, despite extensive debate and pressure from India, Brazil, Japan and others, remains rooted in the geopolitical circumstances of 1945.

Since Security Council reform is moribund, there is no possibility of matching the United States militarily and Brazil, Japan and South Africa have renounced nuclear weapons (thus blocking the route to another prestige indicator), it is unsurprising that the indicators of international status ranking have begun to shift. Here, there appears to be a pattern in which UN bodies and multilateral institutions provide a key arena for status competition and enhancement. In the contemporary international system, one such status indicator is membership of the restricted groups that have attempted to exercise control over economic, financial and energy issues, the G8, G20 and the Major Economies Forum (MEF). The latter is dedicated to climate and energy issues, while the G8 and G20 have, on occasion, attempted to reach high-level political agreements and make declarations that directly 'read across' to UNFCCC meetings. The G8 and G20 are rooted in attempts to manage the Bretton Woods system and both can be seen as responses to global economic crises. The G7 has a continuing existence as a club of developed world finance ministers to which was added, in 1973, an annual summit meeting at heads of government level. The EU Commission was accepted as a participant and, in 1997, the post-Soviet regime in Russia was approved as a member, thus forming the G8.[7]

Emergent powers and their quest for status

Only Japan, along with the EU, has given priority to enhancing its status through climate leadership. For the BASICs, it represents one subsidiary component in a continuing search for appropriate international status. The 'emerging powers' have been defined as 'states that have established themselves as veto players in the international system, but have still not acquired agenda-setting powers (Narlikar, 2013, pp. 561–2). Because they are 'emerging', they are status inconsistent. The reason for making the distinction between self and community ascribed status is that perceived inconsistency, the gap between national perceptions of appropriate dignity and importance and the actual status accorded by the rest of the international community, provides an irritant and a spur to action. This can be associated with a refusal to be socialised into the existing international order and a profound attempt to recast it (Terhalle and Depledge, 2013).

China's re-emergence as a great power provides a much-discussed example. Sinic interpretations of the status of the People's Republic reflect cultural traditions grounded in notions of the exceptional character of Chinese civilisation, where there can be no question as to its rightful place in the surrounding world. This is accompanied by a narrative of national humiliation at the hands of European powers, followed by resurgence. China is a permanent member of the Security Council, but, despite its achievement of WTO membership in 2001 and its hugely significant role in international economic and financial affairs, it did not receive the accolade of being added to the G7. It has also suffered from the loss of reputation associated with the Tiananmen Square Massacre in 1989 and from persistent external attacks on its human rights record and its behaviour in Tibet. In response, the Chinese government has been robust in the assertion of its great power status. This has been observed to be a notable theme in Beijing's climate change policy involving 'a forceful and nationalistic assertion of its sovereign rights' (Christoff, 2010, p. 647) and an emphasis on 'safeguarding national sovereignty and elevating China's prestige in the international community' (Zhang, 2003, p. 82).

India has possessed a self-image as a major power since the days of Nehru, but is an 'underachiever, falling below the threshold for major power status designation' (Volgy et al., 2011, p. 16). Standing between East and West in the Cold War, it fashioned a 'principled' foreign policy of leadership of the Non-Aligned Movement and G77. During the 1970s and 1980s India was a champion of the New International Economic and Information Order, 'common heritage' principles and the sovereignty of developing nations over their resources. Commentators noted the strong ideological and moral character of Indian policy (Mohan, 2010, pp. 136–8). Elements of this remain in the way in which

India champions 'equity' in climate politics and takes the hardest of lines against the EU and Umbrella Group. After the ending of the Cold War and India's rapprochement with the US, there was an evident contradiction between its 'determination to find a way to sit at the top table, while simultaneously deploring it for becoming a concert of great powers' (Mohan, 2010, p. 140). This has also been portrayed in terms of the competition between the two strongest drives of Indian policy, 'the norms of *equity* and global *status*' (Atteridge et al., 2012, p. 74). China is both an ally within the BASIC group and a rival. Indeed, both its campaign to join the Security Council and its attempt to alter the Non-Proliferation Treaty regime rules to accommodate India as a nuclear power were 'visibly resisted' by China (Basiur, 2011, p. 197).

Like India, both Brazil and South Africa are regionally important powers with only 'emergent' status on the world stage. Unlike India, they have both renounced possession of nuclear weapons as an indicator of global status. Brazil only emerged from 21 years of military rule in 1985 and, while in the midst of serious economic problems, was able to play host to the Earth Summit in 1992 – at a moment when South Africa was still an international outcast under the last vestiges of the old apartheid regime. Brazil's bid for membership of the Security Council was unsuccessful and not supported by its Latin American neighbours. Although clearly the dominant country in the region, it is not accepted as a regional leader either by its associates in Mercosur or by the members of the rival ALBA alliance (Herz, 2011, p. 178). The second administration of President Lula, 2007–10, made a concerted attempt to gain major power status where participation in multilateral organisations, and the climate regime in particular, was a central strategy (Herz, 2011; Hochstetler and Viola, 2012). These emergent powers were active within the G77 (South Africa being able to become so after the ending of Apartheid rule) and had begun to challenge the dominance of the G7 countries in the management of the WTO.

As the realisation dawned that the emissions of the developing world would soon outstrip those of the OECD countries, systematic attempts were made to recruit these emergent powers to select international clubs. In 2005, under the British G8 presidency at Gleneagles, an opening was made to create a dialogue between the G8 and five developing countries – Brazil, India, China, South Africa and Mexico (G8+5). The specific objective of a strategy carefully devised by the Blair government (involving meetings with business groups and a climate report by the Hadley Centre) was to utilise the UK presidency of both the G8 and EU to draw the large developing economies into a high-level discussion of the future of the climate change regime post 2012. At the 2007 G8 summit the German presidency attempted to institutionalise this in the 'Heiligendamm process'. It was never a satisfactory relationship, because the five were denied full participation and shared the same peripheral status as a variety of other governments which were, on occasion, invited to attend summits to consider particular agenda items. While France, supported by the United Kingdom, voiced the opinion that it was 'no longer reasonable to meet as eight to solve the big problems of the world', the US and Japan were unconvinced (Reuters, 2008, 5 July).

As part of what was seen as the Bush administration's attempt to bring Beijing and New Delhi into a revised great power concert, the United States created two new climate-related groupings, a partnership within the existing APEC Forum and a new MEF (Drezner, 2007). Both initiatives were directed at undermining EU leadership and the Kyoto Protocol (Liebermann and Schaefer, 2007). Asian countries, including India, China and Japan, were recruited in 2005 to the 'Asia-Pacific Partnership on Clean Development and Climate' that advocated a technology and economic growth-driven alternative to the Kyoto Protocol. The Major Economies Meeting on Energy Security and Climate Change was a broader group, convened in 2007. Its 17 'leader's representatives' were from the G8 plus five, with the addition of Indonesia, Portugal (as EU president) and the UN. The Obama administration re-launched this body in 2009 as The MEF on Energy and Climate. With the addition of Australia, this body convened as a summit meeting in July 2009. It has continued to meet twice per year, but its exclusivity has been compromised by meeting at the level of 'leader's representatives' and by the inclusion of additional participants. In September 2013 there were ten additional countries involved, including the Marshall Islands, Peru and the Democratic Republic of the Congo (MEF, 2013).

G20 and BASIC

The events of 2008–9 catalysed major changes in the international status ordering. The G20 was initially a group (wider than the G7) of finance ministers and central bank governors that had been meeting since 1999. In the context of the deepening international financial crisis of 2008 it convened in November, in Washington, as a summit meeting. This was followed in 2009 by two meetings in London and Pittsburgh. Climate change was on the agenda, but remained a relatively peripheral issue for a body wrestling with the problem of global economic re-structuring and recovery (University of Toronto, 2011). Nevertheless, it has been persuasively argued that, in terms of bringing together those most responsible for climate change and those who could take effective action, the G20 is 'as good as it gets' (Carin and Mehlenbacher, 2010, p. 33). For some time, the relationship between the G8 and G20 was unclear, but the latter has claimed to be 'the premier forum for our international economic co-operation' (G20, 2009). The G20 adds six countries to the G8+5 members: Argentina, Australia, Indonesia, Saudi Arabia, South Korea, Turkey, plus the EU. As with the G8, there was inevitably a politics of exclusion. The comments of the Norwegian Foreign Minister were particularly pointed:

> The G20 is a self-appointed group. Its composition is determined by the major countries and powers. It may be more representative than the G7 or the G8 in which only the richest countries are represented, but it is still arbitrary. We no longer live in the 19th century, a time when major powers redrew the map of the world. No one needs a new Congress of Vienna.
>
> (Støre, 2010)

Large economies which, in strict GDP terms, would be among the global top 20, such as Spain and the Netherlands, remained outside, as did Switzerland and Norway. EU members do, however, have some collective representation. There are five excluded countries with larger economies than South Africa, the smallest G20 member. These include Taiwan, Iran and Venezuela, countries whose economic size is not reflected in overall status due to the opposition of one or other of the major powers. The G20 claims that it is representative, small enough to be effective and includes 'countries and regions of systemic significance' (Townsend, 2010).

For India, and for the other BASICs, the challenge of reconciling their previous positions as G77 members and proponents of 'equity' with their new position as powers with global status, and as direct interlocutors of the United States, was acute – and has yet to be fully resolved. A Chinese diplomat was quoted as saying that 'the *politics* of the negotiations in Copenhagen were "much more important" to China than the climate regime itself' (Terhalle and Depledge, 2013, p. 581). There are different perspectives on the endgame at Copenhagen, but all attest to the significance of status and recognition. Lynas (2009, p. 2) observed an uncontested Chinese superpower, 'its newfound muscular confidence' on striking display:

> The Chinese premier, Wen Jinbao, did not deign to attend the meetings personally, instead sending a second tier official in the country's foreign ministry to sit opposite Obama himself. The diplomatic snub was obvious and brutal, as was the practical implication: several times during the session, the world's most powerful heads of state were forced to wait around as the Chinese delegate went off to make telephone calls to his 'superiors'.
>
> (Lynas, 2009, p. 2)

Terhalle and Depledge (2013, endnote 7) see it as an unfortunate mistake, in which the Chinese leader was under the impression that he was being excluded, while Christoff (2010, p. 647) agrees that China was enigmatic and obstructive during the negotiations and that Wen's absence was interpreted as an insult. China stripped commitments out of the Accord, even for developed countries, and asserted itself as an opponent of the United States, possibly to sustain its leadership of the G77 while simultaneously indicating its place at the top table and maintaining its freedom of action. China's head of delegation concluded that 'the meeting had a positive result ... After negotiations both sides have managed to preserve their bottom line. For the Chinese this was our sovereignty and our national interest' (cited in Christoff, 2010, p. 644).

In June 2013 agreement on joint action to phase out HFCs was announced at the Sunnylands, California, summit meeting between President Obama and his newly appointed Chinese counterpart Xi Jinping. Also at this meeting the Chinese side proposed the notion of a 'new type of great power relationship'. It implies the acceptance of China as an equal partner. By strengthening 'China's view of itself as a recognised and respected power, Xi Jinping is able to foster stronger nationalistic pride under Communist Party of China leadership and gain political capital to consolidate his own power at home' (Li and Xu, 2014). The US side did not endorse the concept, but was prepared to spend a year in detailed discussion of a new joint climate initiative. The fruit of these discussions was announced at a full dress summit held in the Great Hall of the People in Beijing on 12 November 2014 (United States White House, 2014). Behind it lay a convergence of interest and the government of China's reassessment of its energy future, not least on account of the stifling levels of air pollution in Beijing and other cities. The Chinese government's purposes were also served by this clear and symbolic message that the agreement was a manifestation of the 'new type of great power relationship'.

Conclusions

An understanding of status competition and the importance of recognition helps to explain state behaviour, which may not be directly attributable to the pursuit of national energy or economic interests. Academic study has tended to concentrate on the importance of recognition and status inconsistency in the explanation of conflict (Galtung, 1964; Lindemann, 2010). Yet it is also evident that, as well as its role in legitimation and identity creation, the pursuit of prestige and recognition has instrumental functions for the conduct of international business and the advancement of national interests. In this respect there are two ways in which status attribution is relevant. First, it creates the 'expectation of leadership' and legitimises it so that status becomes 'a form of soft power that confers privileges on certain states' (Volgy et al., 2011, p. 10). Second, the enjoyment of high status is said to reduce some of the material costs of efforts to re-structure global order and the 'institutional development necessary for global governance' (ibid.). Such an insight is to be found in classical realist writings. Hans Morgenthau (1967, p. 77) trenchantly denied the idea that only actual material capability would determine outcomes in international politics, for: 'A policy of prestige attains its very triumph when it gives the nation pursuing it such a reputation for power as to enable it to forego the actual employment of power.'

In climate politics, the EU's reputation may yield influence in negotiations and reinforce trust and credibility. This dimension deserves more scrutiny in a leadership literature that tends to emphasise material power and diplomatic skill and expertise in its treatment of 'entrepreneurial' and 'cognitive' leadership (Wurzel and Connelly, 2010, pp. 12–13). Reputation may help to underpin all three. As one Asian delegate confided, 'we know that the EU not only talks the talk but walks the walk.'[8] This may provide part of the explanation as to how the EU was able to mobilise such a wide coalition of developed and developing states in support of its 'roadmap' for a new agreement prior to the negotiation of the Durban Platform in 2011. It was also vital that the EU, unlike other developed Parties, made a pledge to engage in a second commitment period under the Kyoto Protocol. India was left relatively isolated in objecting to the Durban Platform and its willingness to compromise was most likely influenced by concerns about its standing with the rest of G77, which had broadly supported the new Platform. The United States had been subject to similar reputational pressure in 2007 when, after almost a decade of obstruction, it was finally persuaded to accept the Bali Plan of Action. Obviously there was more to this than events at the final Plenary of the 2007 COP, but the record shows how the US delegation moved to avoid reputational damage when it agreed to 'join the consensus' after hasty phone calls to Washington. It had apparently been 'stung by rebuffs from South Africa and Papua New Guinea', the latter urging that the United States should 'get out of the way if you are not willing to lead'. These interventions had inspired 'lengthy applause' by delegates and observers (ENB, 2007, pp. 16, 20).

As Keohane (2010) has pointed out, utilising work on the 'economy of esteem', status-seeking behaviour can be used to promote agreements (Brennan and Pettit, 2004). According to this analysis, there is an 'intangible hand' at work that might usefully be incorporated into the design of institutions. Keohane's suggestions involve establishing a 'high standard of praiseworthy performance' on climate change and then 'evaluating actions by countries and firms according to that standard'. The 'essential mechanism' would then be 'the institution of a number of prominent awards and prizes' (Keohane, 2010, pp. 20–1), something that has recently been set up within UNEP's Minimata Convention.[9]

Arguably, a broader and more significant status competition has always existed in climate politics, but the problem has been to harness it to regime construction. From the early days of the climate regime EU strategy attempted to make use of what might be described as the politics of emulation. By setting ambitious targets and timetables for GHG reductions it was hoped that other Parties would respond in kind and that it would then be possible to 'ratchet up' the overall level of commitment, generating what politicians refer to as 'momentum'. A similar process of emulation seems to operate with pledges of financial support, although, as developing countries complain, these often remain unfulfilled. The appeal is to the competitive desire of Parties to maintain and increase their relative standing. Emulative processes appeared to be under way prior to the Copenhagen COP, with the pre-announcement of emissions targets and funding pledges by the main players, with the EU in the lead offering 20 per cent emissions reductions by 2020 and 30 per cent contingent upon similar pledges by other Parties.[10] However, the outcome was not a success, as EU Commissioner Hedegaard remarked: 'In 2009 China and the US played "after you sir" – none of them really moving as the other party did not move' (quoted in Harvey,

2014, p. 11). Status concerns had ensured that, once it was known that key political leaders including the US President would appear at the conference, others ensured that they would be there as well. One consequence of such a high level of representation and investment of political capital was that the meeting could not be seen to fail. Thus, when it became clear that agreement on text that had been argued over for several years would not be forthcoming, the assembled leaders required an outcome that they could endorse. Such was the genesis of the Copenhagen Accord. Part of the Accord, the invitation to Parties to communicate their national pledges, did provide a mechanism for public competition in emissions reduction ambitions.

After Copenhagen the EU expressed a readiness to raise its level of ambition to a 30 per cent emissions reduction by 2020, but 'not alone and not unless other UNFCCC parties moved rapidly to launch new comprehensive negotiations' (ENB, 2011, p. 29). Subsequently, the EU Commission announced targets for an overall emissions reduction of 40 per cent by 2030, 60 per cent by 2030 and no less than 80 per cent by 2050 (Commission of the European Union, 2011). While rejecting 'strict enforceable rules' and compliance procedures that depress ambition and limit participation, the Chief US negotiator for a 2015 agreement claimed that 'The opposite is true for norms and expectations which countries will want to meet to enhance their global standing and reputation' (Stern, 2013, p. 4). The US–China agreement of 2014 challenges others to follow: expressing the hope that 'by announcing targets now, they can inject momentum into the global climate negotiations and inspire countries to join in coming forward with ambitious actions as soon as possible, preferably in the first quarter of 2015' (US White House, 2014). Once again, there was an appeal to status competition as the driver of agreement.

7
Structural Change and Climate Politics

During the life of the climate regime structural change in the international political system has reflected underlying shifts in the pattern and distribution of economic growth and associated emissions of GHGs. The period 1989–91 has pivotal significance. The ending of the Cold War re-ordered the international political structure and accelerated the processes of economic globalisation, as previously closed economies became enmeshed in a worldwide, market-based system of finance and production. This, in turn, had profound implications for the power structure. In 1992 the United States was the sole remaining superpower, in what was then described as its 'unipolar' moment. In economic scale it was matched only by the EU. In the trade regime and elsewhere it was still possible to portray economic diplomacy in terms of a directorate of two, or perhaps four, advanced industrialised powers. Within a decade, however, China had been admitted to the WTO and, profiting from the decision of many developed world firms to re-locate their production processes to take advantage of its low wage rates, achieved spectacular rates of economic growth, averaging over 9 per cent per annum. Other 'emergent economies' also exhibited high growth rates, leading to perceptions of a new multipolar structure, or even a potential con-dominion of the United States and China (the G2). The events of the 2009 Copenhagen COP were an emblematic demonstration of re-ordered power relationships.

This chapter provides a chronological view of the development of the climate regime within the context of the ongoing changes in the international system, followed by an analysis of some possible structural determinants of power relationships and outcomes in the climate regime.

In political discourse the term 'structure' is an over-worked, yet under-defined, concept. On occasion the adjective 'structural' may merely be a synonym for that which a commentator feels to be profound or significant. Socioeconomic and political structures may be defined as patterns or regularities that can be observed and systematized. Unlike the structural framework of a building or a transport system, they are not manufactured but tend to arise as a consequence of the myriad interactions of the units of a system. Once established, such structures serve to constrain and channel behaviour and to prescribe the choices open to individual actors. The relative significance of structure as a determinant of human behaviour has been a key issue in the social sciences (Giddens, 1984). In IR the 'agent structure debate' has focused on the explanatory power of system-level explanations of state behaviour (Dessler, 1989; Buzan et al., 1993). The best-known variant of structural explanation is to be found in neorealism, as propounded by Waltz (1979). This takes the realist preoccupation with the anarchic character of the international system to extreme lengths by proposing that there is no significant differentiation between states and that all are structurally constrained to behave according to the 'organizing principle' of anarchy. From this perspective, what matters, in terms of outcomes, is the distribution of power across the system.

It is useful to make a distinction between relational and structural power. Relational power is of the everyday variety, where country A can coerce or induce country B to change (or maintain) behaviour through threatening or using armed force, or through utilising its economic muscle. Such tactics are not generally applicable to climate change negotiations, although there have been times when economic inducements and aid funding have been relevant instruments of relational power. Forms of structural power are much more likely to determine the direction of climate diplomacy. As Susan Strange (1988, p. 31) explained, the essence of structural power is 'that the possessor is able to change the range of choices open to others, without apparently putting pressure directly on them to take one decision or to make one choice rather than others'. The question then arises as to which structures are significant. Strange envisaged a quadripartite global structure comprising interrelated political/security, production, knowledge and finance structures.

While realists have been concerned to establish the primacy, and indeed autonomy, of the political/security sphere, most analysts of economic and environmental politics, deriving their inspiration from the liberal or, more particularly, Marxist traditions, would emphasise the significance of economic structures. There is, indeed, an evident relationship between economic performance and achievement of great power status. This is underlined in the literature on 'power transition' and changing international structure (defined in terms of the relative capabilities of the major states) over long historical periods (Organski, 1968; Tammen et al., 2000). In very recent times the (re) emergence of China as an acknowledged great power may be regarded as a consequence of the process of economic globalization. For issues such as climate change, the distribution of economic and technological capabilities and resources across the state system must have relevance because it will determine the extent to which financial inputs and other inducements can be provided. The pattern of emissions will powerfully influence perceptions of which states are indispensable to any agreement on mitigation; but so will the knowledge structure, which will tend to advantage some states over others because they are in a position to frame the issues and set the terms of the debate.

A critical question is whether there can be a satisfactory explanation of outcomes in a particular regime based on overall international power structures, or whether it is the particular structure of that regime that matters. Realist scholars concerned with international political economy have advanced their own proposition on the overall structural conditions under which international cooperation may be possible. This is the 'hegemonic stability' thesis, which requires the leadership of a predominant power to ensure and enforce compliance with international agreements. Here, the absence of hegemonic leadership from an economically and politically pre-eminent United States might provide one possible explanation of the failure to develop a comprehensive and effective climate agreement.

The changing context of the climate regime

The first two decades of international attention to environmental and climate-related issues occurred during an essentially bipolar Cold War system, although East–West relations during the 1970s had been characterised by the relaxation of Cold War tensions known as détente. Relations between developing states, despite aspirations to form an effective non-aligned group, were still coloured by orientation either towards Washington or Moscow. India, for example, relied on Soviet support, while Pakistan looked towards the United States. The pattern was repeated in regional confrontations in Africa, the Middle East and South East Asia. Table 7.1 includes a range of events that occurred during

Table 7.1 The changing context of the climate regime

Date	Contextual events	Climate-related events
1970s	East/West bipolar structure – period of détente. North/South relations focus on NIEO demands at UN	UN Conference on the Human Environment, 1972 World Climate Conference, 1979
1980s	Second Cold War – significant increase in East/West tensions. Marginalisation of the South	Law of the Sea Convention open for signature (1982) but US fails to ratify
1986	Chernobyl nuclear disaster	Rise of Green Parties in Western Europe
Late 1980s	Series of summer heat waves	Montreal Protocol on ozone-depleting substances agreed 1987
IPCC set up 1988
Margaret Thatcher calls for climate treaty at UN, 1989
1989–92	End of Cold War –	Stabilisation of CO_2 emissions at 1990 levels by 2000 under discussion by environment ministers (1989)
1989	Tearing down of Berlin Wall	Second World Climate Conference, IPCC first assessment presented (1990)
1990	German unification	
1991	Dissolution of Warsaw Pact
Outbreak of violent conflict in Yugoslavia	Drafting of Climate Convention by INCs I to V (1991–2)
1992	Dissolution of Soviet Union. Russia assumes UNSC seat	UN Rio Conference on Environment and Development – UNFCC open for signature.
1992	Treaty on European Union formally creates EU
Outbreak of violent conflict in Bosnia-Herzegovina	EU unable to act effectively in Bosnia but expresses aspirations for climate leadership role.
1995	US-brokered Dayton Accords bring cessation of conflict in Bosnia	Berlin CoP proposes measures for adoption, by Annex 1 countries, at the 1997 Kyoto CoP. AGBM meetings
1997–9	Election of Blair government in UK
NATO Kosovo operation
Collapse of WTO Millennium Round at Seattle (1999)	Byrd-Hagel Resolution of US Senate
Kyoto Protocol to the UNFCC opened for signature

2000	Election of President Putin in Russia and US President George W. Bush	Collapse of CoP 6 at the Hague
2001	September 11 terrorist attacks on Twin Towers and the Pentagon prompt 'global war on terror'	Denunciation of US signature of Kyoto Protocol (2001). EU pledges to lead Kyoto development at Cop6 bis and CoP 7 Marrakesh Accords
2003	US-led invasion of Iraq NATO operation in Afghanistan Deadlock in WTO Doha Round	EU capitalises on widespread anti-American sentiment to further Kyoto ratification process, including by the Russian Federation
2004	Accession to EU of eight Central and East European countries.	CoP 10 Buenos Aires
2005	US-India nuclear agreement, G8 Gleneagles Summit	Kyoto Protocol entered into force., ETS set up.
2007–8	Global economic crisis. G20 summit. China becomes guarantor of international financial system.	Economic downturn in the Eurozone undermines EU emissions reduction targets for Copenhagen CoP. Bali CoP 13 agrees twin track road map
2009	Formation of BRIC and BASIC coalitions.	Copenhagen Accord agreed between US and BASICs.
2011		Durban Platform promoted by EU and Cartagena Dialogue partners.

the evolution of the climate regime. Of particular concern are those that are indicative of structural change in the international system.

During the 1970s, the developing countries of the G77 were able to combine in their engagement with North–South issues, for which the UN General Assembly provided an arena. The oil boycotts and price rises imposed by the Arab petroleum-exporting countries at the time of the 1973 war between Egypt, Syria and Israel revealed the economic vulnerability of the OECD countries. This provided a basis for a concerted effort by developing nations to reconstruct the world economy in an NIEO that dominated discussions at the UN General Assembly through the mid-1970s. Related negotiations on comprehensive revisions to the Law of the Sea, at the Third UN Law of the Sea Conference, wound on throughout the decade. General Assembly and UN conference diplomacy tended to advantage the G77, if it could maintain its overwhelming voting majority. G77 alignments at the UN emphasised shared development interests and a common concern to establish the sovereign independence of Southern states in their 'structural conflict' with the United States and former European colonial powers (Krasner, 1985). This helps to explain why developing nations were insistent that the climate issue should come under the auspices of the UN General Assembly.

The period of 1979–80 can be regarded as a turning point in great power relations because it marked the onset of the Second Cold War period of acute tension, deteriorating relations and strategic competition between the East and West. Meanwhile, a new, harder Northern line at the UN saw the termination of North–South negotiations over NIEO. The deterioration of both East–West and North–South relations was, in some measure, the consequence of the election of Prime Minister Thatcher in the United Kingdom and President Reagan in the United States. In retrospect, the final years of the decade acquired further systemic significance because they witnessed the start of market-based economic reform in China under Deng Xiao Ping and the opening up of its previously closed economy to foreign trade and direct investment.

The conclusion of the 1980s saw a profound change in the international structure, as the ending of the Cold War culminated in the collapse of the Soviet Union and the dismemberment of the post-1945 division of Europe. The external political context of the INC negotiations was dramatic – as one participant commented, 'History in Eastern Europe and the Soviet Union was going into overdrive' (Brenton, 1994, p. 85). The terms of what was to become the UN Climate Convention were under active consideration as the Cold War drew to its dramatic conclusion.

In the aftermath of the ending of the Cold War, neorealists and others attempted to grapple with the characteristics of an emergent international power structure. In terms of raw military capabilities, a Russian Federation in economic turmoil remained the only state that was even remotely comparable to the United States, while the latter's main economic rivals were its closest allies. Robert Jervis captured the dilemma that this posed for structural theorists:

> The configuration is so odd that we cannot easily determine the system's polarity. Is it unipolar because the United States is so much stronger than the nearest competitor, bipolar because of the distribution of military resources, tripolar because of the emerging united Europe, or multipolar because of the general dispersion of power?
> (Jervis, 1991/2, p. 42)

A decade later a systematic study that attempted to identify the contemporary great powers found that, although there was a long-term rise in inequality between members of the international system, and the individual strength of even major powers might be in relative decline, collectively the OECD countries continued to exhibit 'overwhelming strength' compared with the remainder of the international system (Volgy and Bailin, 2003, p. 92). The open-ended commitments made by these developed countries in the Climate Convention, and acceptance of the CBDR-RC principle, are puzzling because of the contrast with the imposition of neoliberal orthodoxies elsewhere in North–South relations. One explanation is that the United States and EU were so preoccupied with their struggle over emissions targets that they failed to fully appreciate the significance of other aspects of the Convention (Brenton, 1994, p. 195).

The only sense in which there was a direct and immediate 'read across' to the climate regime from the tumultuous events that ended the Second Cold War, was the inability of the Soviet Union and its successor states, most of which were soon to be classified as economies in transition (EIT), to participate fully in the drawing up of the Climate Convention (UNFCCC). With the Russian and satellite economies in chaos, their contribution to the climate deliberations was generally negative and sometimes aligned with a sceptical United States in opposing targets and obligations. A further, immediate consequence of the ending of the Cold War was the reunification of Germany on 3 October 1990. This is a significant date because 1990 was also the baseline year for measuring emissions reductions under the new UNFCCC. As emissions of highly polluting and inefficient East German plant were now included in the total for the newly unified German state, closing them made it possible to register very substantial reductions (amounting to around 25 per cent). This, along with British reductions resulting from the running down of coal-fired power generation after the Thatcher Government's destruction of UK mining, enabled the EU's 1998 'burden-sharing agreement', which actually allowed some member states to enjoy emissions increases while delivering an 8 per cent EU-wide reduction. The year 1990 remains the EU's favoured baseline.

Alongside these major structural changes, the translation of what had been a largely scientific concern with climate change, the province of experts, to a mainstream political issue in the OECD countries, owes much to some specific environmental events. These included a series of unusually hot summers in the late 1980s, which raised the salience of the climate issue among developed world publics. In addition, the apparent success of the stratospheric ozone negotiations between 1985 and 1990 encouraged the idea that global environmental problems could be tackled through international action, and served to frame the institutional approach to the infinitely more complex matter of climate change. The US government had led on the stratospheric ozone issue against European opposition and, during his election campaign, President Bush senior appeared to adopt a similar role in relation to climate. However, this was soon to alter towards a position of some scepticism, together with resistance to bearing the economic costs of making immediate reductions in fossil fuel emissions. The EU, on the other hand, became committed to an agreement that would stabilise emissions at 1990 levels by 2000. Thus the Intergovernmental Negotiating Committee (INC) and preceding meetings tended to be dominated by transatlantic disagreement over emissions reduction targets and timetables – a pattern that was to persist for the ensuing 20 years.

Creating the Kyoto Protocol – towards EU leadership?

The 1992 Treaty of Maastricht marked the formal creation of the EU, with aspirations to a common foreign policy and commitments to the creation of a single currency. The stated ambition was for the Union to assume leadership in climate diplomacy. As discussed in Chapter 6, this played a significant role in EU identity construction. EU leadership during this period did have some structural basis. At the beginning of the Kyoto process the EU collectively was second only to the United States as an emitter of GHGs. The ending of the Cold War also facilitated the

EU's climate diplomacy, enabling Brussels to 'associate' a whole range of accession states and dependent neighbours with its climate policy positions.

The first Conference of the Parties, held in Berlin in 1995, proposed a process that would elaborate policies and measures within specified time frames for Annex I countries. This was to begin 'without delay' with a view to adopting results in 1997 at COP 3 in Kyoto. The G77 and China were able to avoid any mention of developing country commitments, while JUSSCANZ realised their goals on the acceptance of Joint Implementation (Rowlands, 1995, p. 7). The political dynamics involved changes within both G77 and JUSSCANZ. The G77 divided, as the majority supported AOSIS positions rather than OPEC-led objections to any agreement on developed country commitments (Grubb, 1995, p. 2).[1] In the discussions leading up to Kyoto the EU was prepared to accept US requirements for flexibility in the achievement of climate targets and timetables in return for US adherence to the latter. Achievements on climate were in sharp contrast to the evident weakness of the Union as a significant player in more traditional areas of foreign policy. The focus of transatlantic political interest, at the time of the Berlin Mandate and beyond, was another direct consequence of the ending of the Cold War – the disintegration of former Yugoslavia. Early EU efforts to exercise some control over events in this part of its immediate neighbourhood failed and eventually a reluctant US administration was induced to intervene.

Although Vice President Al Gore signed the Kyoto Protocol, it was clear that Senatorial opposition would preclude US ratification. This left the EU as the main sponsor of the Protocol. The obligation was taken seriously and included a wholesale revision of previous approaches to internal environmental policy in the enthusiastic adoption of emissions trading. The election of George W. Bush in 2000 heralded a dramatic change in US policy and a new opportunity for the EU to assert itself. In March 2001 the Republican administration formally denounced the US signature of the Protocol, provoking a wave of outrage from green NGOs and sympathetic politicians. The EU rose to the challenge and its Stockholm European Council confirmed that the Union would lead the development and ratification of the Protocol.

The events of 11 September later that year were to set in motion the 'global war on terror'. This had some implication for the politics of climate change in so far as it affected transatlantic relations. In 2003 an American-led coalition, including the United Kingdom, invaded Iraq. This divided the EU but also contributed to widespread condemnation of the Bush administration, which dissipated much of the international support that had followed the destruction of the World Trade Center. EU efforts to bring the Kyoto Protocol to fruition became enmeshed in a broader anti-Americanism as the Union was, on occasion, able to present itself as an enlightened alternative to the Bush administration (Vogler and Bretherton, 2006). The EU expended much diplomatic effort in persuading other members of Annex I to ratify the Protocol and was prepared to offer inducements to the Russian Federation. It did so in the face of outright opposition from the United States.

Despite the animosity that came to exist between Washington and Brussels on climate issues, it was still the case that leading member states and the Commission appreciated that an effective climate regime could not be built without the United States. Thus, at the 2005 G8 Gleneagles summit, there was a concerted European attempt to persuade the Bush administration to revise its climate scepticism, followed up when Germany hosted the 2007 Heiligendamm meeting. It remained a primary objective of the EU to find a means to bring the United States into some form of international climate action despite its outright rejection of Kyoto. This might involve sub-federal cooperation with US states and local authorities that were implementing emissions trading schemes. It also meant that a formula had to be found for post-2012 discussions. To this end, the EU sponsored the Bali Plan of Action at the 2007 COP, proposing a twin-track negotiation that would involve the United States in deliberations about the future of the Convention but would allow it to have no dealings at all with the proposed evolution of the reviled Kyoto Protocol.

The EU enlargements of 2004 and 2007 probably mark the zenith of the EU's influence as an actor (Bretherton and Vogler, 2009). Subsequently, it had to cope with the incorporation of economies that enlarged the Single Market but lowered average GDP per capita, while bringing new problems of energy use and dependence. From 2006 continuing energy insecurity represented a significant factor in the deteriorating geostrategic relationship with Russia. In this respect climate policy became entangled with the 'high politics' of European security. East European member states have, for example, resisted collective GHG reduction targets on the grounds that they would, by penalizing the use of domestic coal and shale gas, increase their dependence on and political vulnerability to Russia (Keating, 2014a).

Although it remained an economy on a par only with the United States, it appeared that growth trends signalled inevitable relative decline. By 2009 EU CO_2 emissions represented only 11 per cent of the global total. Above all, the financial collapse of 2007–8, and the ensuing existential crisis of the Eurozone, did serious damage to the self-confidence of the Union and to its external reputation. It had been represented as the quintessential postmodern political form and a source of policy solutions. After 2008 it was more commonplace to regard the EU itself as a problem for the global economy. Great internal effort was expended on its post-2012 pledge to deliver a 20 per cent reduction in emissions by 2020, which, if matched by other developed countries, would be translated into a 30 per cent reduction. The economic downturn in the Eurozone meant that a decline in economic activity would make 20 per cent a less-than-ambitious target. Meanwhile, popular concern with environmental issues was submerged beneath the pressing requirements of rebuilding the European economy. These were the broader economic and political determinants of EU 'failure' at the 2009 Copenhagen COP.

After the disappointment of the Copenhagen COP came a striking revival of EU climate leadership, which tended to contradict predictions based on the EU's relatively weakened structural position. The Union forged alliances with a range of other less-developed and developed Parties in the Cartagena Dialogue (van Schaik, 2012). The outcome was the successful negotiation of the 2011 Durban Platform, initiating the negotiation of a new comprehensive agreement by the end of 2015. This was dependent on EU willingness, in contradistinction to other major Annex I Parties, to participate in a second commitment period of the Kyoto Protocol (ENB, 2011, p. 30). Despite this external success, internal climate policy was beset by difficulties as the ETS was undercut by a falling carbon price and the unwillingness of member states to agree to the necessary reforms to the allowances system. Most humiliating, was the fate of the EU's aviation emissions policy, devised in 2008 as an initiative to overcome the omission of this sector from the Kyoto Protocol. From 2012 airlines using EU airspace were to have been required to buy ETS credits to cover their emissions. However, under strong international pressure and threats of trade sanctions, and beset by internal disunity among member states concerned about the economic implications for themselves, the European Parliament was persuaded to postpone the external operation of the scheme until 2017, pending a decision of the International Civil Aviation Organisation (ICAO) to institute alternative arrangements for taxing aviation emissions (Keating, 2014b).

The Rise of the BRICs and BASICs

The early years of the twenty-first century witnessed the rapid emergence of large developing state economies and, most particularly, the rise to power and prominence of China. The old dominance of the

United States and EU in economic institutions, particularly the WTO, was clearly under challenge. The first evidence of this came at the Seattle WTO ministerial held in 1999 to launch the Millennium trade round. India led the resistance of developing members to attempts to introduce the 'Singapore issues' – liberalising rules of investment, competition and state procurement. China joined the WTO in 2001, in the same year as the launch of a new Doha 'Development Round'. This responded to events in Seattle in two ways, by recognising the demands of developing members and by finding a location that was conveniently inaccessible to anti-globalisation protesters. It was also in 2001 that Jim O'Neill, of Goldman Sachs, coined the term BRICs to give a collective description to Brazil, Russia, India and China – very different economies that were all experiencing rapid growth rates.[2] The Russian Federation, after the chaos of the 1990s, was enjoying an economic revival as a fossil fuel exporter. India, at the head of a coalition of developing countries, resisted Northern plans for the liberalisation of agricultural trade, thus ensuring that the Doha Round stalled. In 2003 at Cancun, and again in 2008 at Geneva, the WTO talks collapsed – an event that could be interpreted as a reflection of the changing global power relations that had eroded the dominance of the OECD economies (Spiegel Online, 2008). The old 'Quad' that had previously dominated WTO discussions (US, EU, Canada and Japan) had already, in 2007, been reformulated into a new version, whereby Brazil and India replaced Canada and Japan. India, rather than China, was regarded as the most obdurate champion of Southern demands. However, in a 2005 demonstration of disconnection between economic and strategic/nuclear diplomacy, the Indians came to an historic nuclear rapprochement with the United States (Narlikar, 2006, 2013).

The global economic crisis of 2007–8 must be regarded as a systemic event of the first importance, unparalleled since the 1930s. Its full ramifications have not yet been revealed, but it reinforced existing trends in dramatic fashion. The EU and US economies teetered on the brink of financial collapse as vast state resources were poured into the rescue of the banking sector. As their economies slipped into recession there was an evident reliance on emerging markets to provide the engine of growth and global recovery. The changed economic circumstances were reflected in widespread agreement on the obsolescence of the G8 and its replacement by the G20 grouping that included the emerging economies. It was this group, meeting at head of government level, which grappled with the global economic crisis in 2008–9. The key outcome of the 2009 London meeting of the G20 was that China was seen to provide the necessary financial stimulus that would rescue the world economy. The, then, British Foreign Secretary David Miliband opined that China had become the 'indispensable power' in stabilising global capitalism:

> The G20 was a very significant coming of age in an international forum for China. If you looked around the 20 people sitting at the table ... what was striking was that when China spoke everybody listened.
>
> (Borger, 2009, p. 2)

Critically, China not only had the capabilities, but also the desire, to play a key role. In climate diplomacy the G20 summit of April 2009 was followed in November by the agreement of India to form the BASIC group and to follow China in offering emissions intensity reductions. This has been portrayed as part of the transition of India to an accepted 'status quo' power that takes part in writing the rules rather than being constrained by them (Mohan, 2010, p. 139).

The Copenhagen COP, scheduled for the end of 2009, had already been ordained as the meeting which would finally agree to a new, comprehensive, post-2012 climate settlement; and preparatory negotiations had been underway since before the onset of the global economic crisis. At the G20 London summit of April 2009, the urgency of the financial situation swept other matters, including climate change, from the agenda. However, the ongoing crisis reinforced demands from OECD countries for the major emerging economies to make a contribution to emissions reductions, alongside those being pledged by developed countries, in advance of the Copenhagen COP. In response, the Chinese government took the initiative to convene what became the BASIC coalition in October 2009, when a memorandum of understanding with India was drawn up. A week before the conference, Brazil and South Africa were invited to join. They then concerted their approach, China and then India having both made offers of reductions in carbon intensity of their economies. At the Copenhagen COP the BASICs, as outlined in Chapter 6, directly engaged with the United States in the writing of the Copenhagen Accord. The fact that the final deal on the Accord was struck between President Obama and the BASIC heads of government dramatised the change that had occurred. A near universal consensus saw this as the expression of a new power constellation that translated underlying structural trends into what amounted to a new great power deal over the future of the climate regime. In the most extreme view, it was an example of the rise of the 'G2', in which the old hegemon came to an arrangement with the Chinese contender. In the aftermath of the 2008 G20 summit a new structural hierarchy, with a G2 of China and the United States at its apex, was widely discussed. The 2009 Copenhagen COP indicated that such a duopoly had not yet emerged. Both China and the United States were locked in extensive economic and monetary interdependence, but the US administration regarded China as a potential military competitor and both sides appeared to prefer to embed their relationship within the wider G20 (Garrett, 2010). However, climate change has been one area of a contested bilateral relationship in which the United States and China have been able to work together. The EU, previously a leader, was seen to have been effectively side-lined. Yet Copenhagen could be represented as a positive and realistic turn, after the prolonged deadlock over Kyoto, born of a more open and cooperative relationship between the major players (Grubb, 2010; Brenton, 2013). It might be argued that the relative exclusion of the rest of the G77 and the EU from the negotiation of the Accord had brought the climate negotiations back into alignment with the structural realities of the global political economy.

After Copenhagen the clear demarcation between Annex I countries and the rest of the world began to dissolve. The annexes themselves remained but the 'firewall' between the developed and underdeveloped, as delimited in 1992, was removed. The Durban Platform of 2011 contained a commitment to produce a new climate agreement, of ambiguous legal status, by 2015 – which would, in highly significant wording, be 'applicable to all'. For the US chief negotiator there would be 'a climate regime whose obligations and expectations would apply to everyone' instead of a Kyoto-based system 'where the reverse was true' (Stern, 2013, p. 3).

Underlying structural trends

In realist accounts of the international structure it is military expenditure and capability that defines great powers. As discussed in Chapter 6, the ranking of great powers remains, in this respect, remarkably stable over the period under consideration. In the post-Cold War system the United States continues to dwarf all other contenders (Figure 7.2). Here, US expenditure is nearly ten times that of its nearest rivals, the United Kingdom and France, with China ranked fourth and India tenth. By 2012, Asian military spending was overtaking the Europeans. To an extent this mirrors the still highly unequal relationship between the

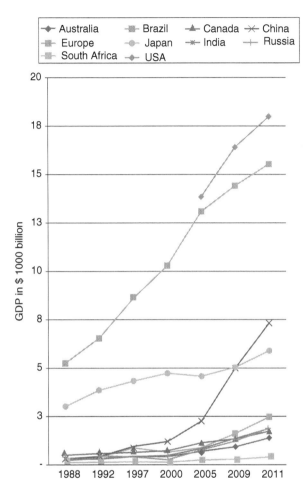

Figure 7.1 GDP growth for major economies
Source: The World Bank (2014) GDP (current US$). Available at: http://data.worldbank.org/indicator/NY.GDP.MKTP.CD?page=4 and http://data.worldbank.org/indicator/NY.GDP.MKTP.CD/countries/EU?display=graph Accessed: 27/06/2014.

gross domestic products of the major developed and developing countries. Other than as a signifier of overall status, the international structure, defined in terms of narrowly strategic criteria, has little relevance to the determination of economic and environmental issues. This had become clear even before the transformation of the international system by the ending of the Cold War. Under conditions of 'complex interdependence' between societies, it was argued, military power was increasingly unusable (Keohane and Nye, 1977). While it remains relevant to attempts to cope with security-related consequences of change in climatic conditions, it is not applicable to negotiations on climate cooperation. A more useful structural perspective, and one which is relevant both to future military balances and the specifics of commercial, environmental and other issue areas, is economic.

If GDP data are used to determine the global economic power structure, then there is a record of substantial change. In the 20 years from 1992 the developed countries' share of world GDP was reduced from approximately 50 per cent to 40 per cent (PricewaterhouseCoopers [PwC], 2012). From 1988 to 2011 Chinese GDP, in terms of current US$, rose from $310 billion to $7,379 billion. Equivalent figures for Brazil are $330 to $2,477; for India $302 to $1,848; for South Africa $115 to $408 and Russia $506 to $1,858 (World Bank, 2013). It is the rate of growth, rather than the absolute size, of the BASIC economies that has fuelled assertions of structural change. While China approximately doubled its share of world GDP between 2000 and 2010, the performance of the other BASICs was much less impressive (Figure 7.1). By 2032 it is predicted that the developing Asian economies will be roughly equal to the developed world, each with a 37 per cent share of global GDP (PwC, 2012). Despite widespread gloom over its economic performance, the EU's share of world GDP between 1988 and 2011 declined only marginally to around 25 per cent, with a total GDP, in 2011, of $16,150 billion. The equivalent US figure was $14,991 billion and that for Japan $5,867 billion (ibid.). Nevertheless, the 1992 division of the world between Annex I and the rest has become difficult to sustain. Four non-Annex I countries have joined the OECD since 1992, including an industrially dynamic South Korea.

The differences in per capita GDP between major OECD economies and most non-Annex I countries, including India, Brazil and South Africa, remain stark. Yet, as US negotiators like to point out, by 2013, 66 non-Annex I countries had higher GDP per capita than the least wealthy Annex I member (Stern, 2013, p. 7). Finally, the gap between the least developed countries and the rest has remained wide and a comparison of the 1992 and 2012 statistics reveals very much the same group of predominantly African and poor Asian countries at the bottom of the global distribution. In this period, the share of non-Asian developing countries in world GDP fell very marginally, to under 24 per cent (PwC, 2012).

Carbon dioxide emissions are central to debates on climate change mitigation and are clearly related to levels of economic activity and

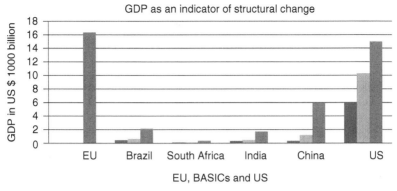
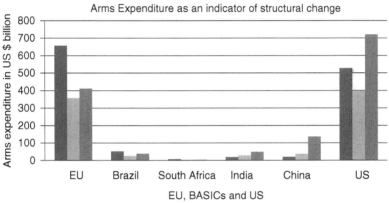
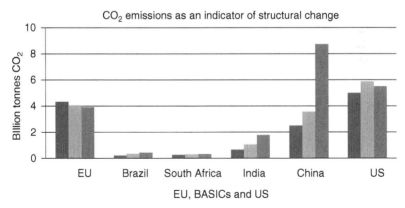

Figure 7.2 Indicators of structural change for EU, BASICs and US 1990, 2000 and 2010

Sources: The World Bank (2014) 'GDP (current US$)'. **(GDP)**
SIPRI (Stockholm International Peace Research Institute) (2014) *'SIPRI Military Expenditure Database 1988–2013'* 'SIPRI Military Expenditure Database' **(ARMS)**. Available at: http://www.sipri.org/research/armaments/milex/milex_database/milex_database Accessed: 27/06/2014.
PBL Netherlands Environmental Assessment Agency (2013) 'Trends in Global CO_2 Emissions, 2013 Report'. Available at: http://edgar.jrc.ec.europa.eu/news_docs/pbl-2013-trends-in-global-co2-emissions-2013-report-1148.pdf Accessed: 03/07/2014 **(CO_2)**.

the phases of industrialisation. Around 1990, the major emitters in both historic and current terms were still the industrialised nations of the global North. US emissions were twice those of China and the EU held second place with around a 19 per cent share of the global total. This provided the physical foundation for the classification of developed world Parties to the Climate Convention in its Annex I; and the rule that theirs was the exclusive responsibility for making immediate reductions. What is striking, here, is that a somewhat different pattern emerges in the period 2000–9. While the record of most countries displays either modest growth, coupled in some cases with a recent decline (the US and EU post 2005), Chinese emissions rise exponentially from around 2000, surpassing those of the US in 2007. Even before this, emissions trends were becoming clear. Thus a 2004 report by the IEA made the important prediction that, by the 2020s, the major part of current emissions would originate from China, India and other developing economies (IEA, 2004). This served as a significant frame for the debates on a post-2012 climate regime. In 1992 non-Annex I countries accounted for 45 per cent of global GHG emissions from energy and industrial uses – by 2013 the figure had risen to around 60 per cent. It is predicted to rise to, perhaps, 68 per cent by 2030 (Stern, 2013, pp. 6–7).

GDP and emissions data provide only the crudest guide to structural shifts in the world economy. Adoption of Susan Strange's (1988) concept of multiple international structures would reveal a highly complex world in which globalised financial markets and transnational business corporations operate to impose structural constraints on the political choices and opportunities of even the largest nation-states. In the production structure there have been huge changes in the location of industry, and related emissions, as developed economies have de-industrialised and factories have moved to Asia. There are also important patterns of cross-ownership and control that do not always involve developed world corporations setting up parts of their supply chains in the developing world. Indian entrepreneurs, for example, own substantial parts of the UK automobile industry and much of EU steel-making capacity. However, the developed economies have generally retained their dominant position in the technological and financial structures, which makes it appropriate for them to offer aid, capacity building and technological support within the UNFCCC. Even here, there have been challenges to this dominance, notably from China. In global monetary affairs the United States and China, which owns a critical amount of US government debt, are structurally interdependent; and in 2008 it was to China, with its large trade-generated currency surpluses, that the rest of the G20 looked for salvation. In renewable energy technologies the Chinese have become the leading producer of solar panels, to the extent that the EU Commission took action against their exports in order to sustain the European industry. For many analysts it is the rapid development and dissemination of renewable and low carbon technologies – such as the much discussed, but undeveloped, Carbon Capture and Storage (CCS) – that hold the key to the future mitigation of emissions, while retaining economic growth (Victor, 2011).

These technologies could revolutionise the global energy production and consumption structure, thus playing a critical role in shaping the definition of national interests discussed in Chapter 4. For a long period the OPEC countries have benefited from a structure based on ever-rising oil consumption, which yielded states such as Saudi Arabia disproportionate influence. Their obstructive position in climate change negotiations rests upon an underlying concern that, ultimately, the structure will shift to their detriment. Indeed, in the last decade the global energy structure has already been significantly altered by the implementation of 'fracking' technology in the United States, which will turn that country into a net energy exporter and, by granting a degree of energy independence, has potentially liberated US policy from previous long-term constraints and imperatives. The situation for China is very different, with its continuing demand for large quantities of externally sourced hydrocarbons. The EU is also enmeshed in an energy supply structure that imposes serious costs and constraints on a broad range of external policies, most particularly in relation to Russia and its 'eastern neighbourhood', which must also bear upon its climate policies. Although energy dependence varies across the 28 states of the Union, there is an excessive reliance on Russian oil and gas which can, as demonstrated by the impact on EU member states of the Ukrainian gas crises of 2006 and 2009, have serious political consequences. This became strikingly evident in the confrontation over Ukraine, the Russian seizure of the Crimea and military support for separatists in the South-East of the country during 2014. Ideally, the Union could devise internal policies of decarbonisation, combined with an efficiently functioning internal energy market, which would augment its energy security while both supporting its leadership ambitions within the UNFCCC and allowing a more robust foreign policy towards the East (Vogler, 2013). The European Commission has proposed such measures in its quest for a new 'Energy Union', but they remain a very long way from realisation.

There are also what may be described as non-material knowledge and ideational structures. In an issue area that is necessarily defined by natural science, the global knowledge structure is highly significant. Around 70 per cent of IPCC authors come from the universities of the developed world. As a Nigerian IPCC author has remarked '... the few of us from developing countries are not able to match their intellectual fire power' and this has subtle effects in terms, for example, of a de-emphasis on the consideration of historic emissions and the related equity issues.[3] The possession of scientific and policy expertise, that derives, in part, from an established position in international scientific networks, is a necessary component of a government's ability to operate effectively within the climate regime. This is an area where NGOs that have access to scientific expertise can play an influential role in advising governments. It also provides a basis for the kind of 'cognitive' leadership that the EU has attempted and, indeed, for initiatives such as the Brazilian proposal discussed in Chapter 5.

The end of the 1980s marked a transition, not only in the structure of the international system, but also a change in what has been described as the international order – its underlying organizing principles and normative assumptions. There was much debate, for example, on the extent to which a new world order, proclaimed by the first Bush administration, had arrived – with Liberal theorists advocating a universal and democratic liberal order championed by the United States (Ikenberry, 2001). In relation to climate change, a key issue has been the extent to which the BASICs could be enmeshed and socialised into the prevailing order. As rising powers, BASIC members have generally been concerned to ensure their full participation; and to avoid incurring mitigation commitments while holding developed nations to their CBDR responsibilities. Behaviour in the climate regime generally accords with that encountered in the WTO, International Monetary Fund (IMF) and elsewhere – in that it displays, since the ending of the Cold War '... greater acceptance of the content of major international governance structures' (Kahler, 2013, p. 718). An unwillingness to exercise leadership or to provide an alternative governance template linked to 'a pragmatic desire for maximum policy discretion to deal with the effects of globalisation' (ibid., p. 715) might also describe their attitude towards the climate regime. Since the terms agreed at Rio have come under increasing challenge as the 'rise of the BASICs' has registered with other Parties, the emerging powers have '... come to see themselves as defenders of the status quo and of established international norms rather than revisionist states seeking to challenge the dominant norms of the system' (Hurrell and Sengupta, 2012, p. 469).

From 2009, however, there is some evidence that the structural situation of the BASICs has begun to lead them into a fuller engagement with the future shape of climate agreement that might be negotiable with the United States and its developed world allies. The economies of the US and China are increasingly interdependent, and energy and climate issues, just as much as monetary ones, can no more be excluded from their bilateral conversations. This emerged at Copenhagen and has more recently been evident in their collaboration on HFC reduction within the Montreal Protocol process.

It is also possible to read the politics of the climate regime as an aspect of a system-wide contest over the nature of the international order. Terhalle and Depledge (2013) argue that a focus on the specifics of climate negotiations may mislead, because the climate negotiations actually reflect disagreements over the international order. Endorsing English School conceptions of related power structures and 'world views', they point to the way in which the Chinese worldview rejects that of the United States. Both are exceptionalist and wish to preserve their sovereignty and internal freedom of action. This involves minimal engagement with the UNFCCC. The climate problem is embedded in a broader political struggle that prevents the formation of the 'grand political bargain that would trigger the paradigmatic change needed to truly set the world's economy on to a low carbon path that can avoid dangerous climate change' (Terhalle and Depledge, 2013, p. 584).

Overall structural power

Realist explanations of international cooperation have emphasised the importance of the international power structure. One structural conclusion that can be drawn from both the strategic and economic evidence is that of the continuing pre-eminence of the United States. This was often referred to as hegemony. In historical and realist literature, hegemonic behaviour denoted attempts by one power to achieve military and political dominance over the other members of the system. Thus, Europe's classical 'balance of power' was seen as a check to attempts at hegemony and a mechanism that would prevent the emergence of a unipolar international structure. The original meaning of the Greek *hegemonia* was rather different. It had the sense of authority or leadership. Realist scholars of international political economy have coined the notion of 'hegemonic stability' (Webb and Krasner, 1989) to denote a distinctive structural theory of the circumstances under which international regimes are created and sustained. In a nutshell, it expresses the ancient idea that the strong make the rules. Cooperation to sustain international economic regimes, which was unlikely to occur through the self-interested behaviour of the participants, required hegemonic leadership from a dominant actor. In its most simplified, premodern version, this would involve a militarily dominant hegemon, but under contemporary conditions this aspect of dominance has a strictly limited application. Nowadays, alongside military strength, one would expect a hegemon to have access to crucial raw materials, control over major sources of capital and to enjoy comparative advantage in goods of high added value. Crucially 'it would be stronger on these dimensions, taken as a whole, than any other country' (Keohane, 1984, pp. 33–4).

During the late 1970s and 1980s much academic effort was expended on considering the apparent loss of hegemony by the United States, the paradigm case being provided by the collapse of international monetary cooperation under the Bretton Woods system of fixed exchange rates that followed the US decision to allow the dollar to float in 1971. Ironically, such discussions of decline focused on the wrong superpower. There has been a renewal of the debate on US decline since 2001 (Zakaria, 2008). Following the costly and inconclusive interventions in Iraq and Afghanistan the Obama administration was clearly unwilling to continue such a high level of external engagement. The economic underpinning of US power had apparently been shaken by extensive de-industrialisation, fiscal irresponsibility and the financial crisis of 2008 (Altman and Haas, 2010). China appeared as a rising competitor and the object of a US strategic 'pivot' towards the Pacific, although Chinese conceptions of re-emergence as a potential super-power were rather more cautious and nuanced (Foot, 2006; Breslin, 2013). Such concerns provided the background to persistent US worries that China, in particular, was growing at the expense of US industry and would benefit from Kyoto-like arrangements that imposed additional energy costs on its competitors.

Given the enormous relative size of the US economy and its continuing hard (and, indeed, soft) power assets, assertions of lost hegemony may be premature. This view is reinforced by remarkable changes in the US energy sector since 2005, where the exploitation of vast reserves of shale gas and 'tight oil' have greatly reduced energy prices, in comparison with competitors, holding out the prospect of energy independence. The geopolitical implications are likely to be extensive, particularly in the Middle East, but they also have direct relevance to climate change policy.[4]

While a plausible case could be made for US hegemonic leadership in the creation of the stratospheric ozone regime in the mid-1980s, there was little likelihood that the United States would be in the vanguard of formulating the international response to climate change, for it had already abandoned its previous leadership of multilateral international endeavours for a narrower pursuit of domestically defined national interests (Falkner, 2005). Although North–South tensions were a constant during the 1980s and 1990s, adherence to the 'common but differentiated responsibilities' principle of the Convention, and the relative economic strength of the United States and EU, meant that the critical climate negotiations were held among these Annex I countries and their allies. The EU was a most unlikely and lop-sided hegemon, but it was able to exercise climate leadership in the absence of the United States. From the initial INC discussions of Article 4.2 of the Convention, through to the Berlin Mandate and its non-ratification of Kyoto, the United States would be better regarded as a dominant veto state, which is another possible hegemonic role. Here it should be recalled that, after 2005, EU climate policy was predicated on the need to find a means to engage its transatlantic partner.

From a power structural perspective, subsequent events illustrate the way in which the long-term equilibrium of a system in which the United States was dominant, and able to determine economic and other rules in negotiation with the EU and other OECD partners, was eroded by changes in production and finance structures, as large parts of the developing world were opened up by the forces of globalisation. The culminating shock to the old structure could be portrayed as the global financial crisis of 2007–8 and the changed 'concert' of powers on display at the Copenhagen COP – just one among a number of indicators of eroded hegemony and a new power structural equilibrium.

There is another type of overall structural explanation that derives from the Marxist tradition. This has long served as a counterpoint to realist notions of structure and hegemony, although drawn from altogether different assumptions about the role of the state and the nature of the international system. A central contention is that climate change is, at root, a problem of global capitalist accumulation. Solutions cannot be found through state governments that are, themselves, the agents of capital. Moreover, international agreements are essentially 'epiphenomenal'. Thus, '... reducing environmental politics to the question of international cooperation and creation of international law, while useful, provides only a limited understanding of the deeper and structural reasons why, despite the flurry of institutional activity over the last forty years, environmental degradation has continued apace' (Newell, 2012, p. 157). From this perspective, analysis of patterns of dominance and dependence in the global economy, and the long-term subordination of most developing nations as commodities exporters and repositories of cheap labour, can provide a convincing structural account of the positions taken up by members of the G77.

Roberts and Parks (2007) have undertaken detailed empirical research into the social and historical determinants of economic and climate inequality in the long-term relations between the developed and developing worlds. Their work provides the essential context for arguments about greater climate equity and access to carbon space. Without arguing that an effective agreement can never be produced, they pinpoint another key component of the recurring problems of the climate regime, which may be placed alongside the lack of hegemonic leadership from the United States. This is a persistent, structural crisis in North–South relations, with an extraordinary imbalance between responsibility for and vulnerability to climate change. The difficulties of the climate regime result from the 'spill-over' of economic development issues into climate diplomacy with, '... unkept aid promises and the onerous requirements of participating in Western-dominated international economic institutions like the IMF and the WTO' (Roberts and Parks, 2007, p. 213).

Issue structure

Overall structural explanations may be parsimonious, but they suffer from a series of limitations when it comes to explaining the specifics of regime change (Keohane and Nye, 1977, pp. 46–9). In particular, they make the assumption that power is 'fungible' across issue areas. The notion that 'power, like water, will find a common level' and 'discrepancies between which states are dominant on one issue and which predominate on others will be eliminated' is, they argue, unlikely to hold under conditions of complex interdependence. Instead there may be a more finely grained 'issue structural' approach to determining how power is wielded in particular issue areas. Within each issue area, this type of explanation 'posits that states will pursue their relatively coherent self-interests and that stronger states in the issue system will dominate weaker ones and determine the rules of the game' (ibid., p. 51).

Issue structural power in the climate regime is, at first sight, likely to accrue to those who can determine mitigation in a significant way. In 2011 six actors accounted for 71 per cent of global carbon dioxide emissions – in descending order, China, the United States, EU, India, Russia and Japan. One might also consider those who are able to provide significant financial inducements, aid for adaptation, technology transfer and technical support. Also there may be others who, like the EU, have both the intent and capability to exercise entrepreneurial and cognitive leadership. This would produce much the same list, with perhaps one or two additions such as the Republic of South Korea.

However, it is not evident from the record that such a relatively restricted group has been able to produce positive outcomes in the climate regime. This is, in part, because their national interests and conceptions of climate justice are significantly at odds. But there are other issue-related institutional factors that may counteract attempts to dominate on the basis of brute material power. The UNFCCC has a complicated and rather specialised issue-related power structure that allows various states and groups, not obviously endowed with major issue-related capabilities, to make demands and to impede agreement. Contrary to experience with the GATT/WTO, or even the stratospheric ozone regime, the climate regime had, from the outset, near universal membership. The very first INC meeting at Chantilly, Virginia, attracted over 100 states. In consequence, Hurrell and Sengupta (2012) observe that the actual power relationships within the regime run in the opposite direction from that which might have been predicted on the basis of underlying economic trends. The 'South', led by China and India, were able to exercise most influence over the negotiations in the early years, in the formation of the Convention and at Kyoto. They were assisted in doing this by the setting of the negotiation under the auspices of the General Assembly, as part of the universalist heritage of the 1972 Stockholm Conference on the Human Environment. This allowed effective coalition politics based on the G77 and garnered the support of Northern NGOs concerned with issues of climate justice. The status of climate change as a specialised environmental issue, and the apparent weakness of many Southern economies in the late 1980s, also helped (ibid., pp. 468–9).

Conclusions

This chapter has shown that structural analysis is important for an understanding of the evolving context of the climate regime. The collapse of the bipolar system and the deepening structures of economic globalisation have combined to facilitate the rise of new players – notably the EU and the BASICs. This is, in turn, indicative of the relative decline of US hegemony. The enduring structural conflict between North and South is also of great significance, both because of the fundamental ethical issues that it raises and of the numerical superiority of the G77 in the UN system.

Study of system-wide material structures, however, can only provide a partially satisfactory explanation of the changing politics of the climate regime. There are clearly anomalous situations where other forms of power have been exercised. The role of AOSIS, an alliance of states with virtually no material basis for influence but which has been disproportionately important in agenda setting, provides a very striking example of how structural weakness can be counteracted, while the Argentines and Dutch have also had a significance beyond what might be expected (Brenton, 2013). The bases of EU leadership also pose a problem for structural analysis. At the outset there was a clear basis for its issue-related power, in the scale of its economy and the extent of its emissions, as long as it was capable of concerting the efforts of its member states operating under shared competence. Copenhagen, however, seemed to indicate a decisive loss of EU leadership. Indeed, with its rather brutal display of great power pre-eminence, Copenhagen appeared to represent a dramatic break with the old climate regime and the emergence of a new concert. The peculiarities of issue structural power within the UNFCCC, however, appear to have moderated this development. Thus the EU, aided by its associates in the Cartagena Dialogue, was able to re-assert its leading position by negotiating the deal between the Umbrella Group, the BASICs and the rest of the G77 that allowed agreement of the 2011 Durban Platform.

8
Conclusion

Nadrev Saño, head of the Philippines delegation, whose country had witnessed successive climate change-driven 'natural' disasters in 2012 and 2013, called the UNFCCC process an 'annual carbon-intensive gathering of useless frequent flyers'. He was announcing his protest fast for the duration of the 2013 Warsaw COP. Later, several hundred NGO representatives staged a walkout in order to register their frustration with its lack of progress (ENB, 2013 p. 30). Such views of the UNFCCC regime are widely shared. Over the years it has become complex and highly institutionalised – even, perhaps, a site of ritualised behaviour.

After a relatively brisk start in terms of ratification and entry into force, the UNFCCC proceeded quickly to the drafting, in outline, of the Kyoto Protocol. This then took more than seven years to enter into force, by which time consideration of action beyond 2012 was required. It took until that year to agree to a limited second phase of the Protocol, which did not include major emitters beyond the EU and Australia and was still lacking the necessary ratifications in 2014. In what can be regarded as significant regime change, principles and norms finally shifted in terms of overcoming an absolutely strict division of the responsibilities of the Parties between Annex I and the rest. Emissions reduction commitments were not to be part of a new agreement, being replaced by the concept of nationally determined 'contributions'. This represented a retreat from the 'top down' targets and timetables model of the Kyoto Protocol, which was probably the necessary price to be paid for a comprehensive agreement on mitigation alongside a greatly expanded role for adaptation funding. In a major shift in the evolution of the regime, the establishment of parity between adaptation and mitigation was demanded by many developing Parties. After the disappointment of hopes for an effective post-2012 regime, the point at which a future comprehensive agreement would come into force has been pushed out to 2020.

This concluding chapter seeks, first, to evaluate the achievements and failings of the climate regime. They are not self-evident, for the regime is open to the charge that its functions are essentially political and symbolic, where governments, in Tony Evans's (1998) memorable phrase, wish to be seen 'doing something without doing anything'. Sometimes it is difficult to avoid the impression that there has been a good deal of 'kicking the can down the road' in order to delay potentially difficult and costly decisions, while reaping the shorter-term political benefits of achieving some form of agreement. Susan Strange (1983, p. 342) made a distinction between regimes that serve the strategic interests of a dominant state, those that are 'adaptive' to cope with changed conditions in the global political economy and those that are purely 'symbolic'. As discussed in Chapter 6, participation in the regime does have important symbolic functions; and governments would suffer serious reputational damage if they simply abandoned it. Nevertheless, it is not merely a forum for the dissemination of the rhetoric of sustainable development; rather it deserves to be evaluated in terms of its stated objectives as an 'adaptive' and operational regime. This means establishing the regime's effectiveness.

This book has also attempted to consider some aspects of climate politics that are not always part of mainstream functional analysis, but which may be significant for future regime building. To this end, some of the findings in previous chapters are reviewed in terms of their relevance to suggestions for improving the design and operation of the climate regime and for the general study of international environmental politics.

The effectiveness of the climate regime

There are various ways of considering regime performance (Victor et al., 1998), including those adopted by IPCC authors. Their performance indicators for the climate regime comprise environmental effectiveness, aggregate economic performance, distributional impacts and institutional feasibility (IPCC, 2014a, p. 58). For the analysis in this chapter, a different but partially overlapping scheme is adopted, which divides discussion into outputs, outcomes and impacts.

The outputs of the climate regime comprise the internal legal architecture of the regime, the nature of commitments and the extent to which they can be enforced. A regime output of interest to a student of IR is the location of authority. Here, the key issue is the anarchic character of the system, the absence of proper governance and the extent to which authority can be transferred to a higher level, thereby limiting the independence of states and internationalising norms and rules. Typically, as Underdahl (1980) has shown, in his 'law of the least ambitious program' international environmental agreements tend towards the lowest common denominator of agreement between state Parties that will only accept decisions by unanimity. This is not always necessarily the case (Hovi and Sprinz, 2006), as the internal EU politics of environmental decision-making can demonstrate. For 'transfer of authority', effectiveness might be measured in terms of movement along a hypothetical continuum between unbridled national independence and a central world government.

Regimes are designed to have outcomes that modify behaviour. For the climate regime this would most obviously be behaviour that leads to GHG emissions and the destruction of sinks. Thus, a key dimension of any evaluation must be the impact of regime rules on observed behaviour deemed to be under the regulatory control of the Parties. Does the regime actually curb damaging behaviour and encourage technology transfer or facilitate carbon trading? Does it encourage and support adaptation activities? In more subtle ways, continuing involvement in the regime may alter the perceptions and the behaviour of governments and other participants to the extent that beneficial, institutionalised learning processes may occur.

Ultimately, effectiveness must refer to observed environmental impacts in relation to stabilisation of levels of atmospheric GHGs and the avoidance of dangerous climate change. This is the area on which most research effort has been expended and where evaluation is both critical and extremely problematic, in terms of establishing complex causal relationships and anticipated effects.

Outputs

The UNFCCC has developed an impressive legal architecture and has near universal membership of 196 Parties. The initial 'framework' nature of the regime indicated its function of assessing scientific evidence and national actions and data. If the IPCC is considered as part of the wider regime, then its research efforts have been unprecedented at the international level. The decision-makers at the UNFCCC, however, have simply failed to respond or to consider the adequacy of their commitments. Nevertheless, the regime has had real success in implementing and reviewing 'national communications' and greenhouse gas inventories, the main obligation of Parties under the 1992 Convention. While there are serious capacity problems for less-developed Parties, a great deal of relevant experience has been acquired as the system has continuously evolved and tightened its methodologies. This has major implications for the monitoring of compliance with the terms of any future climate arrangement, particularly if it comes to be based on 'pledge and review' principles.

Kyoto constituted an ambitious 'control Protocol' containing legally binding commitments for Annex B (virtually synonymous with Annex I) Parties. The regime was weakened by the withdrawal of the United States and, subsequently, by the refusal of other developed countries to engage in a second commitment period. One of the arguments put forward by advocates of the Kyoto Protocol, when charged with its rather limited success in emissions reduction, is that its real longer-term significance is in the construction of novel legal and institutional arrangements. A key element is to be found in the operation of the flexibility mechanisms, described as 'multilayered governance', in building 'institutional capacity in developing countries' (IPCC, 2014a, p. 58). Additionally there are international accounting procedures, offset mechanisms for developed countries and monitoring, facilitation and enforcement arrangements for the CDM.

To this extent, the Kyoto Protocol has been effective in providing a basis of experience and rules for future attempts to create global carbon markets. Unfortunately, however, the regime has not proceeded along a path of increasingly effective institutionalisation. In designing a new post-2020 agreement there has been an evident retreat from strong legal commitments and compliance arrangements. Uncertainties and differences between the Parties are reflected in the ambiguous wording of the *Ad hoc* Working Group on Durban Platform's (ADP) mandate – to develop 'a protocol, another legal instrument or an agreed outcome with legal force under the Convention' (Decision 1/CP.17).

If we evaluate effectiveness as the transfer of authority, a similar pattern emerges. Kyoto represented a 'top-down' set of 'targets and timetables', expressed as specified commitments in the form of QELROs. These have now been abandoned for a Second Commitment Period by all but the EU and a few other Parties. The likely replacement is a much more voluntary regime that substitutes the language of 'contribution' for 'commitment' and relies on 'bottom up' national pledges rather than agreed international targets. The Copenhagen Accord, with its collection of individual pledges, represents a minimal level of international agreement. Indeed, it has been described as creating a 'faux' or even a 'zombie' regime.[1]

In response to these inadequacies, there have been indications of a retreat from the idea of a meaningful, comprehensive global agreement towards decentralised, partial initiatives on tackling 'black carbon', promoting energy efficiency and/or alternative technologies (Hulme, 2010; Nordhaus and Shellenberger, 2010; Prins et al., 2010). Alternative suggestions highlight the individual 'building blocks' of a global climate regime, which could include forestry or regional trading and technology initiatives. In order to maintain legitimacy, they would ultimately be integrated into a globally agreed framework. Its advocates claim that given 'prevailing interests and power structures, a functioning framework for climate governance is unlikely to be constructed all at once in a top-down fashion'. Instead there should be an ongoing political process that 'seeks to create trust between nations and build climate governance step-by-step out of several regime elements' (Falkner et al., 2010, p. 258).

One of the dangers in pursuing a new global deal at the UNFCCC is that effective rules will be traded for a comprehensive agreement. The character and extent of commonly agreed rules is a critical determinant of whether the ADP negotiations in 2015 will produce an output that deserves to be called a new regime. Certainly the Parties will have to raise their level of ambition to something that makes sense in terms of the 'gap to close' (discussed below) and that represents more than a statement of what they would achieve anyway, in the absence of any agreement. A successful regime, utilising some form of 'pledge and review' mechanism, would require the provision of a robust, even intrusive, means of ensuring that Parties' pledges are transparent and their implementation verifiable.

Outcomes

The immediate way in which the climate regime functions to reduce GHG emissions is to modify behaviour that is deemed to be regulated by participating governments. Thus the Kyoto Protocol was directly responsible for the creation of the EU's ETS, countless other policy changes by governments and actions by businesses in response to the flexibility mechanisms. Carbon prices have not been stable, and there are serious concerns over the fraudulent use of the mechanisms, but it is the argument of proponents of the Kyoto system that, meagre as the achieved emissions reductions may have been, what has been created with such effort provides an institutional foundation on which to build. A long and intensive learning process has been 'an important dynamic' for the regime 'The sheer amount of information and analysis passing through the regime in support of the negotiations is massive' (Depledge and Yamin, 2009, p. 441). This has been most marked in the development of novel market arrangements for the CDM and Joint Implementation (JI). Similar justifications would apply to the belated development of forestry instruments under REDD+ and adaptation measures. Adaptation represents a potentially huge cost for the international community but it is difficult to assess the scale of the problem, in contrast to the quantified 'emissions gap' scenarios that are considered below. UNFCCC Secretariat and World Bank estimates put the possible cost of adaptation at a round figure of $100 billion per year while the 2013 level of multilateral adaptation funding was only $3.9 billion (UNEP, 2013, p. 2).

A general movement towards the regulation of GHG emissions has also occurred. In a much-quoted finding, the IPCCC Fifth Assessment Report observed that 'there has been a considerable increase in national and sub-national mitigation plans and strategies ... In 2012, 67 per cent of global GHG emissions were subject to national legislation or strategies versus 45 per cent in 2007' (IPCC, 2014, p. 28). There is also evidence of some increase in the global adoption of renewable energy sources over the same period (Figure 8.1). One of the aspirations for a climate regime is that it should send out signals on anticipated changes that will be internalised by decision-makers. It is impossible, however, to establish a direct causal link between the regime and these modifications in governmental and corporate behaviour. They are, in any event, far too limited to bring about the changes in energy production and use that would be likely to meet the requirements of mitigation scenarios sufficient to stabilise concentrations of GHGs at safe levels. This would require major increases in investment in renewables and very substantial divestment from the 'stranded assets' of the fossil fuel industries, coupled with even larger efforts in 'energy efficiency across sectors' (IPCC, 2014, p. 28).

Impacts

The ultimate test of a regime is to be found in its physical ability to solve the environmental problems for which it was created. The leading example of such effectiveness is provided by the Montreal Protocol, which can now be associated with observable reductions in the concentrations of CFC and HCFC gases in the atmosphere. Because these are also

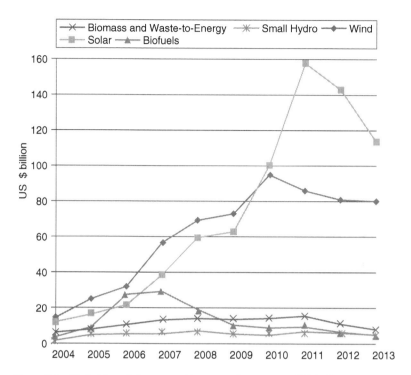

Figure 8.1 Global trends in renewable energy investment 2004–13
Source: Frankfurt School-UNEP Centre/BNEF (2014). 'Global Trends in Renewable Energy Investment 2014: Key Findings' (Executive Summary). Available at: http://fs-unep-centre.org/sites/default/files/attachments/14008nef_visual_12_key_findings.pdf Accessed: 11/07/2014.

significant GHGs, the Montreal Protocol has, in fact, succeeded in the avoidance of GHG emissions that would otherwise have occurred – some 10gt of CO_2 per annum. It thus transpires that the emissions reductions achieved by the Montreal Protocol have been five times greater than those scheduled under the first commitment period of the Kyoto Protocol (IPCC, 2014a).[2] Unfortunately, the CFCs and HCFCs controlled under Montreal are being increasingly replaced by an 'ozone safe' but highly climate-damaging class of chemicals – the HFCs. Left unchecked, their emission could undo all the GHG emissions savings achieved by the Montreal Protocol, producing, by 2050, global warming effects roughly equivalent to those of global transport emissions. In June 2013 President Obama and Chinese President Xi Jinping agreed to collaborate on an existing US proposal to 'phase down' HFCs through the Montreal Protocol Machinery (UNEP, 2013). This agreement is beyond the remit of the UNFCCC regime and illustrates another significant development in global climate politics, in that – given the relative stasis of the UNFCCC – it has been bypassed or outflanked by a whole range of climate- and forestry-related initiatives including, for example, the Regional Greenhouse Gas and Western Climate Initiatives at a sub-federal level in the United States.

For the UNFCCC regime, effectiveness may be measured in terms of the scale of the physical effort required to achieve the 'ultimate objective' of the Convention, which is 'stabilization of greenhouse gases in the atmosphere at a level that would prevent dangerous anthropocentric interference with the climate system' (Art. 2). UNEP provides a synthesis of relevant research using various emissions scenarios in its 'emissions gap' reports. The climate problem and its solution is represented in terms of the 'emissions gap to close'. This is calculated by, first, estimating the level of GHG emissions in 2020 that would likely be required to put the world on a 'least cost' pathway to stabilization at the 2 °C threshold – a median estimate of 44gt of carbon dioxide equivalent ($GtCO_2e$) per annum. Then a comparison is made with best estimates of actual recent emissions – a median figure for 2010 of 48.8 $GtCO_2e$ *per annum* – and their likely growth. Thus, 2010 emissions are 14 per cent above the desired 2020 level and, by then, are likely to rise to 59 $GtCO_2e$ under 'business as usual' assumptions – revealing a gap, if nothing were to be done, of around 17 $GtCO_2e$ (UNEP, 2013, pp. 3–5). The analysis in the report reveals that the regime has not been totally ineffective and might prove, under certain assumptions, capable of narrowing the emissions gap significantly by 2020. If all the current pledges made by Parties – including those of the Kyoto second commitment period, Annex I pledges recorded in 2010 and NAMAs announced by 55 developing countries – are taken together and fully implemented, then the effect by 2020 would be to reduce the emissions gap by more than half, to 8 $GtCO_2e$. Achieving this would require that conditional pledges are fulfilled and that rule changes are strictly adhered to.[3]

The possibility exists that the gap might be entirely closed through additional action by the Parties to increase energy efficiency, remove fuel subsidies, reduce short-lived GHGs such as methane and promote renewables. This may be unrealistic, but it remains important to approach as closely as possible the 'low cost emissions pathway' by 2020. Otherwise, the costs of future action to avoid the 2 °C threshold become increasingly onerous as, for example, carbon-intensive technologies are 'locked in' and the 1.5 °C threshold required by small island developing states is breached. It will also determine the scale of the task that a new climate agreement operative in 2020 will have to confront. If the emissions gap is closed, then the global emissions targets to be met in 2025 and 2030 will be of the order of 40 and 35 $GtCO_2e$ respectively (UNEP, 2013, p. xiii). If not, the mountain to climb will be a great deal steeper and more hazardous (Figure 8.2).

Conclusions on the impact effectiveness of the current regime are by no means entirely negative, but it is clear that a very substantial task lies ahead. The outline of the likely architecture of the post-2020 regime, to be finalised at the end of 2015, is emerging – but it is unlikely to satisfy the onerous requirements of coordinating international action towards a decarbonised world and coping with the rising demand for adaptation.

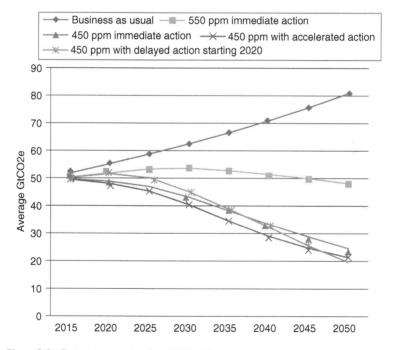

Figure 8.2 Emissions gap to close 2015–50
Source: OECD (2012) OECD environmental outlook to 2050, the consequence of inaction. Available at: http://dx.doi.org/10.1787/9789264122246-en Accessed: 03/07/2014.

The politics of climate change and the future of the regime

In the Introduction to this book there was reference to the functional character of much of the literature in the field, driven by an urgent and understandable quest for effective international solutions to the climate problem. To complement such approaches, the political dimensions of creating a climate regime have been the focus of the current study. The Conclusion to this book provides an appropriate place to review its findings, such as they are, in relation to the larger functional task of building a more effective climate regime.

Chapter 2 considered the framing of the climate problem at the international level. It is generally acknowledged to be a particularly 'wicked' or 'malign' problem reaching into almost every corner of human society and economy; and its drivers lie at the heart of modern carbon-fuelled industrial economies. In this, it is the antithesis of the stratospheric ozone-depletion problem, which involves a specific class of artificial chemicals for which there are, generally, cost-effective substitutes. Nevertheless, experience with the Montreal Protocol meant that the climate problem was institutionally framed in a similar way, but with the critical exception that there was no equivalent attempt to phase out and control the trade in, for example, coal. Such an agreement could probably never have been negotiated, given the vast national and corporate vested interests involved, but it would have provided a more immediately effective framing of the climate issue area. It can also be argued that the framing of the problem was insufficiently broad or deep, given its focus on national GHG emissions sources and sinks. The IPCC Fifth Assessment Report states a common understanding that:

> Globally, economic and population growth continue to be the most important drivers of increases in CO_2 emissions, both have outpaced increases in energy efficiency. The contribution of population growth has remained steady in the decade 2000–2010, but the increase in the contribution of economic growth has increased sharply in the same period and, worryingly, previous trends in the reduction of coal use have been reversed.
> (IPCC, 2014, p. 8)

The reasons why the international definition of the climate issue area excluded what were known to be the principal drivers of the enhanced greenhouse effect were also discussed in Chapter 2. They reflect the difficulties of working within a system dedicated to the extension of economic globalisation and reluctance to address the 'toxic' question of population growth. Above all, it should be emphasised that the definition of the issue area does not so much represent scientific consensus, but rather what is acceptable to governments. It was not for nothing that the original agreement on the UNFCCC was formulated in an inter*governmental* Committee (INC) and that even global scientific research is processed through an Inter*governmental* Panel (IPCC).

It was also argued that there was a relationship between the international framings of the climate problem and the observed fragmentation of issue areas; and attempts at climate governance in an increasingly crowded 'regime complex'. In part, fragmentation has been a political choice involving 'forum shopping' and attempts to divert attention away from the central UNFCCC regime, targets and timetables and CBDR-RC. In the case of the G8, G20, MEF and APEC initiatives, the results have been unimpressive in terms of concrete actions.[4] Elsewhere, however, fragmentation may have some positive benefits in relation to the idea of a 'building blocks' approach referred to above.

Chapter 4 underlined the difficulty of reconciling national interests and alignments in a new climate agreement. The context is one of sharply rising energy demand, and insecurity. The design of the climate regime, emphasising energy-related CO_2 emissions, has meant that major states now take up positions that reflect their energy security requirements alongside concerns over competitiveness in global markets. The economic downturn from 2007 appears to have sharpened the tendency to place short-run economic considerations in the forefront of national policy – as illustrated by the defection of major developed states from a second commitment period of the Kyoto Protocol. China and other emerging economies have also been careful not to undertake commitments that could circumscribe their rate of economic growth. The extent to which domestic politics limit the possibility of international action was also evident, most notably in the United States, where the near certainty of Congressional opposition reduces the President's options to those which can be accomplished under existing legislation.

In the analysis, national interests were represented as a balance between perceived economic costs of mitigation actions and vulnerability. While the costs of action are overestimated, and would be relatively small if early action on climate was undertaken, vulnerability to climate change impacts tends to be discounted; a process accentuated by the wave of orchestrated climate change scepticism that has been a staple of political debates within much of the developed world over the last decade.[5] Future climate-related events may alter the calculus of vulnerability to produce a heightened perception of mutual 'common fate' interdependence among governments. Such a sense has been an underlying characteristic of other commons regimes that have been successfully developed, but under much more advantageous circumstances (Vogler, 2000, p. 224).

There are some contrary examples indicating at least the possibility of different positions on the climate issue. Pre-eminent is the EU, that, despite its own economic difficulties and the setback of the Copenhagen COP, has been able to articulate a progressive position that stresses the 'co-benefits' of climate change action and the potential of decarbonisation policy for both enhanced security and employment growth. This approach has, in part, been shared by the AILAC coalition in Latin America. The building of the 2011 Durban Platform, on the basis of a dialogue between states spanning all the main negotiating groups, shows the way in which skilful diplomacy is capable of aligning conceptions of national climate interest.

In Chapter 5, the justice claims that are so central to the North–South dimension of climate politics were analysed. The CBDR-RC principle and the division of Parties into Annexes remain a part of the regime that the Umbrella Group have been unable to erase. Exactly what part they will play under a new comprehensive, but differentiated, agreement remains unclear. On the one hand, the original 1992 'firewall' is no longer tenable, as current developing world emissions exceed those of the OECD countries. On the other hand, attempts to introduce a 'level playing field' approach to fairness in respect of current emissions alone are also politically and ethically unsupportable. The finances available to the Green Climate Fund, and the extent to which adaptation is treated in as serious a manner as mitigation, will also be crucial to an agreement that needs to be endorsed by the bulk of the international community. Communitarian and cosmopolitan thinking helps to distinguish between the various proposals, but fails to provide a just solution that could command general approval among the Parties. Probably the most hopeful sign is the development of 'no harm' approaches and the evolution of new responsibilities for richer countries in the provision of climate-related aid and adaptation funding, and in emerging conceptions of 'loss and damage' compensation.

Prestige-seeking and demands for recognition are ever-present in international politics. Chapter 6 reviewed some of the ways in which status competition, and sensitivity to diplomatic affront and to perceived violations of national sovereignty, have figured in the life of the climate regime. They can be obstructive under certain circumstances – witness ALBA's intervention at the Copenhagen COP and the obstacles to transparency and verification that are erected in the name of the proper recognition of national sovereignty. Yet status competition can also make a contribution to agreement. The politics of identity construction explains part of the motivation behind the EU's climate activism, and possibly that of other Parties as well. The reputational concern of governments, both domestically and in relation to their international peers, constitutes an important resource continually exploited by those campaigning groups that gather in large numbers at climate conferences and also by governments that wish to persuade others of the advantages of not standing in the way of a consensus.

The politics of emulation has been an important feature of climate diplomacy. EU policy has, over 20 years, sought to challenge other Parties to match its 'targets and timetables'. Under the Copenhagen Accord and the Durban Platform, Parties were urged to submit their intended mitigation contributions to the UNFCCC Secretariat, which publishes them in tabular form. This provides some basis for a competition, although the variation in baselines and the nature of the contributions makes comparison difficult. At key points there have been attempts to generate momentum for this emulative process by a 'ramping up' of the reputational stakes. This has involved raising the political level of representation at Climate COPs and holding 'climate summits' under the auspices of the UN Secretary General. It represents a conscious attempt to exert reputational pressure on governments by taking climate negotiations decisively out of the realm of 'low' functional politics. Such a process was evidently under way in the months preceding the 2015 Paris COP. The events at Copenhagen in 2009 provide a salutary case study of the inherent dangers of this approach. Failure to agree to concrete measures at the highest levels can depress expectations that have been artificially raised, thus publicising the weakness of the regime.

Reputation also has a role in the implementation of agreements. The Kyoto Protocol developed complex facilitation and compliance arrangements based on denial of access to the benefits of the 'mechanisms', but it is unlikely that formal enforcement will be part of a new climate agreement. The climate regime is hardly unique in this respect, for there are few direct sanctions to enforce the rules in a system of sovereign states. If an agreement is not based on a reciprocal relationship of mutual interest then it is frequently reliant on 'horizontal enforcement', based on a concern for good standing and the maintenance of reputation. These are powerful incentives towards compliance – even when short-run material interests may dictate otherwise (Henkin, 1979). This is why clarity in the publication and evaluation of national contributions is so important for a 2015 agreement.

Structural analysis of the kind undertaken in Chapter 7 leads to a consideration of who should be at the table in negotiations, but also who holds a veto. In climate negotiations there is a clearly defined group of six Parties, China, the United States, EU, India, Russia and Japan, responsible for around 70 per cent of global GHG emissions. China, by far the largest emitter, and the United States between them account for over 40 per cent and their participation is, therefore, the *sine qua non* of an effective agreement. This represents a change from the initial period of the climate regime, when the United States was the largest emitter as well as the pre-eminent economic and military power.

Realist theorists have proposed that regimes will be created and maintained by the authority of a hegemon. The earliest systematic analysis of the international politics of climate change came to the conclusion that 'hegemonic stability' theory failed to account for the emergence of the regime (Paterson, 1996, pp. 94–101). The US hegemon played the role of veto state, leaving the EU to exercise leadership. From a realist perspective, if there cannot be hegemonic leadership, then a concert of powerful players will be required to drive agreement through. Its members will tend to overlap with that of the small group of 'essential' nations whose participation is required on the basis of their emissions. The formation of the Copenhagen Accord can be portrayed as such a concert of powers in action (Brenton, 2013). Within this group, the relationship between the two largest emitters has had increasing significance as the structure of the international system shifted in the early years of the twenty-first century. Up until 2013 there seemed little prospect of Sino-American collaboration, and their confrontation over the climate issue was modelled as a game of 'chicken' in which both sides chose to avoid making commitments to emissions reduction (Ward, 1993). In this context, it is easy to understand the significance that has been accorded to the November 2014 bilateral climate announcement by Presidents Obama and Xi Jinping, as a precursor to an agreement in Paris.

There is a similarity between realist advocacy of a great power concert and proposals for a 'club' or 'minilateral' approach to economic and environmental negotiations that relies, not so much on power dynamics, but on a simplification of the geometry of negotiations and control over the issues. The intent is to delineate a group small enough to negotiate effectively but consistent with maximum control over the relevant issues. Analogies are often drawn to the successful formation of the General Agreement on Tariffs and Trade, which expanded from a small influential core of developed trading nations. David Victor (2011, p. 255) argues that the club approach would require '... a few sympathetic governments – ideally the largest beneficiaries of an effective climate club, such as the EU, US, China, India and Brazil (much smaller countries that are nimble and strategic might also play a role)'. The problem facing a club approach is, of course, that negotiations continue to be conducted through the UNFCCC, which has institutionalised a set of practices and understandings (including its failure to agree to rules of procedure) that have tended to democratise discussions along the lines of the UN General Assembly – which, it will be recalled, sponsored the Convention in the first place. There is a contradiction between, on the one hand, the desire to be part of an effective and powerful club and, on the other, reputational damage to national standing and influence within the UN system. It is a dilemma that is particularly acute for the BASIC countries with respect to their position in the G77; and one that was clearly evident in the climate regime after 2009 as many governments reacted angrily to their apparent exclusion. The Cartagena Dialogue helped to overcome some of the resentment, but demands for an inclusive and 'Party-driven' process remain strong. The Co-Chairs of the ADP have been guided by the imperative of including the submissions of all Parties in a broadly based negotiation process, but the result has been a proliferation of alternative 'elements for a draft negotiating text' running to 1,888 lines (CP/2014/L.14), from which a final agreement will have to be crafted.

The practicalities of producing a 2015 agreement highlight the intersection of the demands for procedural justice, the contradictory reputational concerns of the Parties and the requirement for a suitably small group that combines negotiating efficiency with a recognition of key structural determinants. The latter include not only national shares of the global emissions total but also less intangible calculations of the distribution of power within the regime. Historical responsibilities for, and vulnerability to, climatic changes will also have to be represented, if the final text of the agreement is not only to be effective but also legitimate.[6]

Theorising the International relations of climate change

The intention in writing this book was to place the climate regime in its international context, to understand how its development was moulded by broader trends in the international system and how the motivations of actors, beyond the more obvious pursuit of material national interests, could be understood. Climate change is physically related to many other areas of environmental concern, but it has achieved much greater political salience in the international system. It might be justifiable to treat other instances of international environmental cooperation in isolation, as independent functionally oriented negotiations. As E. H. Carr (1939, p. 102) observed, not all the business between states is 'political', for instance the maintenance of postal services or the suppression of epidemics. 'But as soon as an issue arises which involves or is thought to involve the power of one state in relation to another, the matter becomes political'. This is evidently so for the climate change issue but, to the extent that it is also true for other issues, the approach developed here may prove to have broader relevance.

In attempting to understand the international politics of climate change, this book has drawn inspiration from a variety of well-established theoretical approaches to the study of IR. It is not written from the vantage point of any one of them. As the discipline of IR has moved in diverse and frequently incompatible directions, a satisfactory general theory seems further away now than ever. In combining a number of approaches it has been assumed that, in the absence of a single master theory, these various theoretical orientations have analytical value in illuminating the whole from different perspectives. Such exercises are often said to provide alternative theoretical 'lenses' for viewing and interpreting events. Some of the theorists quoted may seem decidedly ancient if not canonical, Carr, Wight, Morgenthau and Wolfers plus other works that date back to the 1970s and 1980s. They represent an important part of what is still the mainstream of IR theorising, which this author would characterise as a set of ongoing debates rather than a single cumulative theoretical enterprise. These timeless debates, between realist and liberal conceptions of interest or between normative cosmopolitan and communitarian positions, provide a useful way of organising and investigating the politics of climate change. Many scholars still follow in the realist or liberal internationalist tradition, while others, for example from the English School, have been attempting to apply a classical approach to the relatively novel subject of the 'greening of international society' (Falkner, 2012). Prestige, and the need for recognition, were acknowledged by early thinkers as important ends of state policy. A recent revival of theoretical interest in such topics, and their connection to constructivist understandings of identity politics, provide another example of how long-standing IR approaches may be revived and reconstituted in ways that are directly relevant to the understanding of contemporary climate politics.

Whereas it is commonplace to analyse the national economic interests of the parties to international environmental negotiations, considerations of power, prestige and recognition and the political functions of negotiations, beyond the achievement of their stated objectives, also require consideration. A full understanding of events within the framework of the UN or within the wider setting of interstate politics will necessarily involve a conjunction of actors' pursuit of material interests, prestige and, of course, justice. Therefore, this book has adopted a tripartite approach to the understanding of agency in interstate relations. Chapters 4, 5 and 6 cover the interests, alignments and motives of the Parties to the Climate Convention. Chapter 7 follows established practice in shifting the 'level of analysis' to the structure of the system. Orthodox realist analysis focuses on the overall power structure of the system but it is not immediately evident that this, rather crude, pattern of relations is reproduced within the climate regime. Instead, in analysing specific outcomes within climate negotiations, it may be more accurate to consider the extent to which specific 'issue structural' power resources provide a more satisfactory explanation of outcomes.

In the climate issue area other features of the system will come into play – obviously the global structures of production, finance and consumption, but also the knowledge and technology structures. The key to understanding is often located in the complex interaction among, for example, the power structure, national material interests, status competition and justice claims. The international politics of the 2009 Copenhagen COP provides a good example. It is difficult to specify, with any precision, the exact way in which the interaction occurs and the weighting to be given to the various factors. The relationship between power structure, justice and moral concerns has been a perennial theme in IR although often downplayed in realist writing. To quote E. H. Carr again, 'In practice we know that peaceful change can only be achieved through a compromise between the utopian conception of a common feeling of right and the realist conception of a mechanical adjustment to a changed equilibrium of forces' (Carr, 1939, p. 233). This old dialectic is reproduced in novel ways in the climate regime, as the moral fervour of climate activists and arguments over the nature of differentiation interact with the 'changed equilibrium of forces' in the early twenty-first century.

One criticism that certainly applies to many of the diverse theoretical positions in contemporary IR is that they cannot be legitimately combined because they rest upon different ontological foundations. They may simply be 'incommensurable', in the sense that they are incapable of being judged, measured or considered comparatively without common factors or units of the same dimension. This does not apply to realism and liberalism because they share an ontological position that recognises a world of states. For this they have been much criticised, but they are appropriate to a book that attempts to comprehend the ways in which the interstate system has both constructed and confronted the climate change problem. The underlying assumption is that the 'international' continues to matter. This rests upon the empirical observation that sovereign states remain the primary locus of public identity and legitimate authority and that they are unlikely to be supplanted by alternative political forms within any time frame relevant to the management of global climate change. In the context of the study of global climate governance this has been subject to extensive challenge, not least because of the successive failures of the UNFCCC to live up to expectations.

Studies of global environmental governance have focused on the significance of a growing mass of sub-state activity aiming to decarbonise economic activity; the regulatory role of the private sector in setting its own standards and innovative transnational links between, for example, major cities. Political mobilisation has also begun to concentrate on actions by individuals and investors without governmental intervention. The worldwide movement to encourage 'divestment' in fossil fuel resources, that can be effectively framed as 'stranded assets', rather than useful sources of long-term revenue, provides a good example.

There is a danger that interstate action and the UNFCCC process will simply be seen as, at best, a minor distraction and, at worst, a dangerous irrelevance to genuine global climate governance. In practice, both the formal interstate regime and myriad local and transnational governance activities are in a complementary relationship. As mentioned in the early part of this chapter, there is already some observable correlation between the formal interstate regime and what is occurring in the rest of the regime complex. Advocates of transnational climate governance have noted that the Kyoto Protocol '... opened new opportunities and incentives for transnational governance' (Andonova et al., 2009, p. 58). Our understanding of the ways in which developments in the interstate regime actually do provide the parameters within which long-term investment decisions are taken remains sketchy and possibly over-optimistic. This is a question worthy of further systematic research. At the same time, activist pressure for decarbonisation may contribute to the incentives for governments to offer more ambitious pledges in an international 'bottom up' regime of the type that has emerged since the Copenhagen Accord.

As ever, the US and European debate about the place of international cooperation in global climate governance tends to prioritise mitigation. Adaptation had barely a mention in the 1992 Framework Convention, but as the damaging impact of mean temperature increases, even below the 2 °C threshold, has become increasingly evident, so the question of adaptation and loss and damage compensation has risen in importance. State action appears to be the only realistic way in which the very major transfers of funds and technology, that will be required to achieve satisfactory adaptation, can be achieved. From a developing world perspective this means that the UNFCCC framework is of great and continuing significance. Unlike many other climate-related fora, the Convention under the auspices of the UN General Assembly remains the body within which developing countries are likely to have the greatest influence on critical adaptation issues.

The discussion of problem framings in Chapter 2, and of the relationship between the politics of prestige and identity in Chapter 6, drew upon constructivist theory. Here there is a potential problem of incommensurability both in terms of ontology and epistemology. While realists and liberals operate on the basis of assumptions of rational choice and pursue causal explanation, the constructivist credo emphasises constitutive understandings of the world. The extent to which they may legitimately be combined represents one of the major and continuing controversies of contemporary IR theory (Wendt, 1994; Smith and Owens, 2008). In this book, constructivist ideas have been used alongside those of more orthodox theory as they seem particularly appropriate to the task of understanding how the issue areas that determine regime creation are framed. The framing of the climate issue area, staked out in 1992, is remarkable in that it fails to address the specific sources of increasing GHG emissions in the production and burning of fossil fuels. It also excludes rival, and what many would regard as more accurate, socioeconomic framings of the climate problem. This continues to be a matter of great significance and it is relatively easy to see that the framings and exclusions of the climate regime are in accord with broader understandings, concerning the primacy of markets and the promotion of globalised economic processes, which predominated after the end of the Cold War. Such ideas probably achieved their greatest acceptance around the moment of the creation of the UNFCCC – indeed essentially neoliberal text is to be found in the Convention. Thus, within the climate regime, there is an evident connection between the restricted framing of the climate problem and the economic interests of leading state governments and their corporate clients. Yet this is not the whole story, for it is difficult to fit the acceptance of CBDR-RC with a simple assertion of the hegemony of ideas supporting a specific constellation of economic interests, or to understand the occasionally contradictory behaviour of actors seeking prestige as well as material advantage.

Governments have continued to prioritise their immediate economic interests within a narrow conception of their national interest and energy security. For developing states, which are not responsible for the historical accumulation of GHGs in the atmosphere and which face urgent issues of poverty eradication and the provision of basic services, this is understandable. The projected impact of climate change is generally much greater for them than for Annex I countries, but there is still a tendency for most governments to discount the mounting evidence of vulnerability and, thus, to neglect the ultimate welfare of their citizens. It is only AOSIS and some of the more vulnerable developing state governments that have *per force* grasped the true gravity of the situation and fully internalised the need for immediate action. This is not an impossible task because the technologies are available and the requirements for stabilisation of the climate system within a defined time window are well understood.

The problem is that the available instruments to coordinate the required national actions reside within the current architecture of the UNFCCC regime. This in turn is inevitably a product of the wider international system. There is a mismatch between the prevailing assumptions and practices of the international political system and the pace of change in the physical environment. Competition for short-run economic advantage and struggles over prestige and recognition continue while the whole basis of the system is threatened by climate change. If, in the absence of effective national and international action to mitigate and adapt, the threshold of dangerous climate change is crossed, then the consequences will be severe. As discussed in Chapter 2, there is already speculation as to the security and geostrategic consequences of an altered climate. Yet the consequences could go well beyond this. They would very likely include a re-ordering of the structure of the global political economy threatening, for example, the mega-cities located at sea level, which are now a major feature of the twin trajectories of globalisation and urbanisation. They could most certainly disrupt the assumptions on which the smooth upward growth curve for the GDP of emergent world powers is based. In future, our attention may no longer focus on the implications of the current international system for climate negotiations, but rather on the manifold and extreme impact of alterations in the climate for the fundamental structures of the world system itself.

Notes

1 Introduction

1. Julian Saurin's (1996) essay still provides an excellent statement of this view.
2. Readers unfamiliar with these theoretical orientations in IR may choose to consult Baylis, Smith and Owens eds. (2013) which in this and previous editions provides clear outlines of realist, liberal, cosmopolitan, communitarian and constructivist accounts of international relations. A more focused discussion of the relevance of IR theory to global environmental politics may be found in Kutting ed. (2011) and in the first part of Harris ed. (2014).

2 Framing and Fragmentation

1. Goffmann's work was focused upon individual representations in interaction. He apparently understood, but did not explore, the institutional embeddedness of framing rules (Goffmann, 1986, pp. xv–xvi).
2. A US official report on the likely state of the world in 2000 noted a number of potential scenarios. The most accurate, with the benefit of hindsight, was that for 'moderate warming' – forecasting an increase of up to 0.6°C by the end of the century (US Council on Environmental Quality and Department of State, 1981, p. 60). There is discussion in the Report of the socioeconomic consequences of what were then known as 'hard' versus 'soft' energy paths, between coal, oil and nuclear on the one hand and renewables on the other. The main concern was with the demonstrable effects of energy choices in terms of land use, nuclear waste and acidification. Nonetheless the combustion of fossil fuels still gave rise to '...concern that a gradual, irreversible and probably dangerous change in the world's climate could occur over the next century as a result of the "greenhouse effect"' (Ibid., p. 353).
3. An element of the Convention that had already been established in the various international conferences and in the significant first IPCC report was that such an agreement should cover greenhouse gas 'sinks' as well as 'sources', the former being a useful synonym for the sensitive North–South issue of forestry (Brenton, 1994).
4. In June 2013 US President Obama and Chinese President Xi Jinping agreed to collaborate on an existing US proposal to 'phase down' HFCs through the Montreal Protocol Machinery (UNEP, 2013). Not the least of the advantages of this move for the United States is that it avoids action under the UNFCCC and, most important, does not necessitate reference to the US Senate for new treaty approval or legislation.
5. The problem of GHGs was only addressed in July 2011 when the IMO agreed to amendments to MARPOL Annex VI (ch.4) which introduce a system of 'technical measures' energy efficiency standards for new ships and a template for ship efficiency management projected to lead to a 20 per cent

reduction in emissions per tonne/km in the 2020s and 30 per cent efficiency improvements in new ships by 2030 (International Chamber of Shipping, 2012, p. 5). The issue of maritime emissions trading has proved a difficult one for the industry, as well as any related attempt to introduce CBDR principles – it being pointed out that through the procedure of 'flagging out' only 35 per cent of the existing fleet is registered in Annex I countries (Ibid., p. 3). Discussions are taking place on possible market-based measures linked to a fuel compensation fund for developing countries.
6. The industry argues that technical improvements and increased fuel efficiency will provide an answer, but this will be a very long way off given the long replacement times for aircraft and the continuing upward trend in numbers of flights.
7. The ICAO aimed to introduce new regulations by 2016 to be implemented by 2020. The arguments within the organisation involved criticism of EU attempts at 'extraterritorial taxation', US fears over setting a precedent and Chinese and Indian objections that developing countries' airlines should not be under the same obligations as those of the developed world (Keating, 2013).
8. In 2004 Sir David King, the then UK government's chief scientific adviser, made this controversial case by arguing that, after the unusually hot summer of 2003, the casualties attributable to alterations in the climate were likely to outnumber those resulting from terrorist attacks. Terrorism was, at the time, placed at the head of the national security agenda and the intent was to suggest that climate change should be treated with similar urgency. This approach was summarily rejected by Downing Street and the prevailing definition of global security threats as a triptych of terrorism, failed states and weapons of mass destruction remained in place.
9. Interview with German diplomat, Bonn, April 2013.
10. The Anthropocene concept was first used by Paul Crutzen (2002), who was one of the scientists sharing the 1995 Nobel prize for work on stratospheric ozone-depleting chemicals. The classification of the current epoch as the Anthropocene, succeeding the Holocene, is widely discussed but not fully established among the scientific establishment.
11. UN population growth projections vary widely according to assumptions on fertility rates. Constant fertility rates would yield a world population of 11.1 billion in 2050 and 28.6 billion in 2120. More realistic projections range between 8.3 and 10.8 billion for 2050 and an actual reduction to 6.75 billion in 2100 (UN, 2012, *World population Prospects: the 2012 Revision*).

3 The UNFCCC Regime

1. The Convention also recognises, but does not define, 'least developed states'. Turkey refused to sign the Convention in 1992 because it was listed as a developed state. There have only been a handful of European additions to Annex I, including: Croatia, Cyprus, Malta, Liechtenstein, Monaco, Slovakia and Slovenia. In 2011, while Cyprus was added to Annex I, Russia, opposed by Saudi Arabia, raised the question of the periodic review of the Annexes (ENB, 2011, p. 4). Annex B, which lists Parties making commitments under the Kyoto Protocol, mirrors Annex I of the Convention except that Kazakhstan

is included, the United States decided not to ratify and Canada withdrew in December 2012. For the second round of the Kyoto Protocol, agreed at Doha in 2012, the only significant Annex I members to have expressed willingness to make new emission reduction commitments are members of the EU. UNFCCC Article 4.2(f) contained a provision for the review of Annexes I and II before the end of 1998, but with 'the approval of the Party concerned'.
2. They provided something around which the rest of the G77 could unite in the face of OPEC members who expressed their opposition to any agreement (Grubb, 1995; Rowlands, 1995, p. 7).
3. Surplus AAUs were a contentious issue in negotiations for a second Kyoto commitment period. The EU currently holds the largest surplus of AAUs but is unlikely to use them (UNEP, 2013, p. 7).
4. The Working Group on the future of the Kyoto Protocol (AWG-KP) was set up at the Montreal COP at the end of 2005, while the working group on 'Long Term Co-operative Action' under the Convention (AWG-LCA) was the creation of the 2007 Bali COP. Both continued to meet until they were wound up at Doha in 2012. Their replacement was the Working Group on the Durban Platform (AWG-DP) which was tasked with negotiating a new climate agreement in time for the Paris COP scheduled for the end of 2015.
5. Alongside the EU, with its 28 member states, the other Annex I Parties to have done so are Australia, Belarus, Kazakhstan, Monaco, Norway, Switzerland and Ukraine. Notable absentees are Japan and the Russian Federation.
6. In the draft decision the proposal was to replace the term 'commitments' with 'contributions', without prejudice to the legal nature of the contributions and substituting the wording 'parties in a position to do so' with 'parties ready to do so' (ENB, 2013, p. 14).
7. They were to communicate detailed information on their policies and measures and projected anthropogenic emissions by sources and removal by sinks (Art.4.1; Art.12). Reporting obligations were always differentiated according to the CBDR-RC principle. Developed countries were required to produce more extensive information within six months of entry into force (Art.4,2b). Developing countries, on the other hand, were allowed three years, or to make their initial communication dependent upon the receipt of financial aid. Least developed countries were permitted to report at their own discretion (Art.12). The role of the Conference of the Parties in reviewing the obligations of the Parties, scientific evidence, the implementation of the Convention and, to these ends, the promotion of comparable methodologies, was also established in broad outline under Article 7.
8. Interview with a Malaysian official, Bonn, April 2013.
9. The EU has proposed a step-wise approach, in which national pledges should be ambitious, transparent, quantified, comparable and verifiable and proposed pledges should be subject to robust international assessment before being inscribed in the 2015 agreement (EU, 2013). In the US view there should also be 'an opportunity for others to analyse and pose clarifying questions regarding...contributions before they are finalized' (United States Government, 2014).
10. By 2013 the GEF had funded $4 billion for mitigation actions and $603.4 million for adaptation through its Least Developed Countries Fund and $200.2 million for adaptation through its Special Climate Change Fund (UNFCCC, 2013, p. 17).

11. REDD+ was part of the Copenhagen Accord and subsequent agreement at Cancun. REDD originally involved reducing emissions from deforestation and forest degradation. After Cancun 2010 new elements to increase environmental integrity were added, including the conservation and enhancement of forest carbon sinks and sustainable management. For this reason REDD is now referred to as REDD plus (REDD+).
12. The ADP like other Conference bodies has Co-Chairs, one from an Annex I and one from a non-Annex I Party. The role of a contact group is to forward agreed text to the Plenary for approval (UNFCCC, 2011, p. 15).
13. The item to which the Russian Federation, Belarus and Ukraine objected, disallowing their right to sell pre-existing surplus AAUs in the second Kyoto Commitment Period, was gavelled through by the Chair (ENB, 2012, p. 27). At the following meeting of the SBI in June 2013 no progress was made for two weeks because of an agenda dispute initiated by Russia, Belarus and the Ukraine over this procedural issue (ENB, 2013, p. 2).
14. Interview with a developing Party delegate, Bonn, April 2013. The role of NGOs was clearly helpful 'they even draft text for us in negotiations'.

4 Interests and Alignments

1. A parsimonious unit level explanation is provided, centring upon the domestic factors that shape external policy preferences on the assumption that 'each country is a self-interested actor that rationally seeks wealth and power by comparing the costs and benefits of alternative courses of action' (Sprinz and Vaahtoranta, 1994, p. 78). The two key variables are the degree of national ecological vulnerability and the economic costs of abatement. While other factors are acknowledged, they suggest that 'different degrees of ecological vulnerability and economic capacity explain much of the cross-national variance found in support for international environmental regulation' (ibid., p. 79). Low abatement costs and levels of vulnerability imply that a country will be a 'bystander', while low costs and high vulnerability will make a country play the role of 'pusher'. Those with high costs and low vulnerability will be 'draggers', while those which are high on both scales can operate as 'intermediaries' (ibid., p. 81).
2. The Lisbon Treaty (Treaty on the Functioning of the European Union [TFEU]), that entered into force in 2009, can be read as giving the Union (in effect the Commission) the right to conduct negotiations where the conclusion of an agreement is necessary to the achievement of the objectives of the Treaty (Art.216 (1) TFEU). Under Art.191 combating climate change is one of these objectives. In early 2010, there was a dispute over whether the Commission or the Council would submit information required by the Copenhagen Accord. Furthermore, there was an argument over who should represent the Union in the upcoming UNEP negotiations for a convention on controlling mercury. At its first meeting the Union was unable to present a position and this was widely seen as a test case for the much more significant climate negotiations.
3. Earlier attempts, in 1992, to develop a carbon tax as a basis for the European position were blocked in Brussels by very extensive industrial lobbying (Skjaerseth, 1994).

4. Byrd-Hagel Resolution, 105th Congress, 1st Session, S.RES.98. The resolution stated that the United States should not be a signatory to any new agreement '...by Annex I Parties, unless the Protocol or other agreement also mandates new specific and scheduled commitments to limit or reduce greenhouse gas emissions of Developing Country Parties within the same compliance period or (B) would result in serious harm to the economy of the United States'.
5. Transcript of a briefing given by the UK G8 Sherpa, Monday 4 July 2005. Final communiqués for G8 'summits' are always negotiated well in advance of the actual meeting, by the aptly named 'sherpas'. The May draft had stated unambiguously 'We know that our world is warming', but by June the phrase had been enclosed in square brackets (*Times*, 18 June 2005). By the final communiqué it had disappeared. It was replaced by the US preferred wording under which climate change was a potential long-term challenge. References to scientific evidence for actual climate change and its anthropogenic causes were also excised.
6. In Canada the Harper government was responsible for the loss of 5,000 funded posts in environmental and climate science over a five-year period (CBC News, 10 January 2014). In Australia the Abbot government cut climate science funding by 70 per cent and closed down the Independent Climate Commission and Climate Change Authority (*Nature*, 2014, 'Australian budget cuts hit science jobs' vol. 511(7507).
7. In the event, the Abbot government's attempt was unsuccessful at the November G20 meeting. President Obama made public references to Australia's climate problems and insisted, along with EU governments, that, in a summit overshadowed by the events in Ukraine, a climate change commitment should be part of the communiqué. He also used the occasion to announce a $3bn US contribution to the GCF (Reuters, 2014, 'US, EU override Australia to put climate change on G20 agenda' 15 November).
8. In 2013 China's per capita carbon emissions were calculated at 7.2 tonnes and the EU at 6.8 tonnes per annum. The comparable US figure was 16.5 tonnes (Global Carbon Project, 2014; McGrath, 2014).
9. It partly explains, for example, why the United States and China have been able to agree on HFC reduction under the Montreal Protocol. This does not require the passage of new legislation, but falls within the existing powers of the Environmental Protection Agency.

5 The Pursuit of Justice

1. There are many excellent discussions of ethical theory in IR and the problem of justice. See, for example, Frost (2006), Brown (1992) and Shapcott (2010) Hedley Bull (1977) distinguishes order from justice. Order is the 'thin' communitarian society of states, preserving coexistence. Justice is the 'thicker' solidarism, more ethically ambitious and more demanding of states.
2. Communitarianism is perhaps more than realism. It asks for an 'eggbox' international relations, where states are cushioned from each other's calculating advances, yet brought closer together so that they can cooperate. Buzan (2004) offers a relevant pluralist/solidarist distinction. Pluralism does not go beyond the basics, and seeks to preserve a 'liveable international order'. Climate change would, of course, violate such a liveable international order.

3. Cosmopolitan conceptions of climate change are explored extensively in the essays included in Harris (2011). For cosmopolitans in the Kantian tradition there may be a green categorical imperative, whereby humans are ends and not means, applicable to climate politics. My thanks to Duncan Weaver for making this point.
4. Pluralism is noted for its concern for state consent. There can be cooperation only when states give their consent; deliberately exercising their sovereignty for the international common good.

6 Recognition and Prestige

1. There are analogies, here, to the arms control regimes of the Cold War in which, in a system of institutionalised distrust, only that which could be verified could be agreed.
2. The Copenhagen Accord produced a negotiated piece of text that addressed the MRV issue in relation to non-Annex I NAMAs: 'international communication and analysis under clearly defined guidelines will ensure that national sovereignty is respected'. Only mitigation actions that receive external support would be subject to international MRV.
3. According to a US diplomatic cable, the heart of Cuba's complaint about Copenhagen was not the substance 'but rather the process, in particular the fact that Cuba, Venezuela and Bolivia were not involved in the negotiations', but the protest at Copenhagen was also used as a 'much needed distraction' from the government's domestic failures (US Embassy Cable, 2010, 7 January).
4. There is an informative piece of research by Karlsson et al. (2011) which provides evidence of the perceptions of participants at the Poznan COP 14 in 2008. The EU and China were the actors most widely regarded as leaders: EU leadership was recognised by '...respondents from Asia, Europe, North America and Oceania, whereas respondents form Africa and Latin America to a higher degree see China rather than the EU as leaders' (ibid., p. 98).
5. Remarks by then President of the Commission, Barroso, to the European Foreign Affairs Conference, 17 April 2010 (author's notes).
6. The Nagoya Protocol to the Convention on Biological Diversity sets out a framework for access to, and the fair and equitable sharing of, benefits arising from their use. UNEP's Minimata Convention on Mercury seeks to ban production and trade in products containing mercury from 2020, to control emissions from coal-fired power stations and limit its use in gold mining.
7. Subsequently, in 2014, punitive action was taken by the Western members to exclude Russia over its actions in annexing the Crimea.
8. Interview with a Malaysian delegate, Bonn, April 2013.
9. There is a working example in the 2013 Minimata Convention on Mercury. In order to encourage monetary and other contributions to the negotiation of an implementing Protocol, UNEP has created a 'Mercury Club' in which state Parties, international organizations and even individuals are presented with gold, silver or bronze awards according to the level of their contribution (UNEP Mercury Partnership, 2014).
10. During the Copenhagen conference itself, and in its aftermath, there was continuing debate over the adequacy of the EU's headline target of a 20 per cent

emissions reduction. It became increasingly clear that, in comparison to the pledges announced by other Parties under the Copenhagen Accord, the EU's 20 per cent lacked ambition, but also that the economic downturn would, in itself, serve to achieve a substantial part of this reduction. Environmental NGOs, the Commission and some member states, including the United Kingdom, pushed for a target of 30 per cent, regardless of any action by other Parties. Advocates argued that, far from putting economic growth at risk, a 30 per cent commitment would encourage investment in low carbon technologies that would inject dynamism into a flat European economy.

7 Structural Change and Climate Politics

1. JUSSCANZ agreement was helped by division of industrial and fossil fuel lobbies, which had wielded negative influence at INC 11, in particular by out-manoeuvring coal interests (Grubb, 1995, p. 7).
2. The BRICs were to become an actual coalition which, at Russian invitation, held its first meeting in June 2009. In 2010 South Africa was invited to join. At Durban in 2013 the BRICs began to develop a programme to call for the redistribution of votes at the IMF and to set up their own development bank.
3. According to Chuks Okereke, a lead author for the Fifth Assessment Report, 'The IPCC is a genuine effort to reflect the underlying science but in the end it does not fully capture the views of developing countries, because the overwhelming majority are from the developed world... they come with loads of secretaries and helpers, they bring their PH.D students along and the few of us from developing countries are not able to match the intellectual fire power that comes from them'. Quoted in McGrath M. (2014) 'IPCCC scientists accused of "marginalising poor nations" ' BBC News, 12 April.
4. For a review of these changes see Hastings et al. (2013). They write: 'It is no small irony that realists, whose IR paradigms stress the importance of material power, have been found in the forefront of those arguing for America's relative decline, often overlooking important changes in the material base of American power that pose a significant challenge to the heart of the declinist case.'

8 Conclusion

1. These comments were made by participants at the Jubilee Symposium of the Centre for Climate Policy and Science Research, Linkoping University, held at Norrkoping, 9 May 2014.
2. Even before the implementation of the Kyoto Protocol, the decrease in aggregate emissions of Annex I countries actually exceeded the voluntary target of returning to 1990 levels by 2000. The Kyoto Protocol requires emissions reductions from Annex B countries (virtually equivalent to Annex I in the Convention). 2011 figures for aggregate emissions reductions under the Protocol are in the range 8.5–13 per cent below 1990 levels, greater than the 5.2 per cent reduction target in the Protocol but mainly achieved by large reductions, which might have occurred anyway, in the Economies In Transition (UNFCC, 2014a, pp. 58–9).

3. This is a reference to the changes made by the regime at the Durban and Doha COPs, which tightened the accounting rules on the use of LULUCF and restricted the use of emissions credits. Improvements could also be made through the avoidance of double counting. Four cases are put forward where reductions from BAU leave a gap varying between 8 and 12 $GtCO_2e$, ranging from full implementation of pledges under strict rules to only unconditional pledges under lenient rules (UNEP, 2013, p. xvii). The ADP discusses the question of current emissions under the heading of 'Workstream 2'.
4. The very limited and partial activities of the G8, G20 and MEF on issues such as fossil fuel subsidies are evaluated by IPCCC (2014a, p. 48).
5. According to IPCC, stabilising to atmospheric concentrations of 445ppm CO_2eq by 2100 would 'entail losses in global consumption – not including benefits of reduced climate change as well as co-benefits and adverse side-effects of mitigation – of 1–4 per cent (median 1.7 per cent) in 2030' (IPCC, 2014a, p. 15).
6. Robyn Eckersley (2012) has proposed a novel way of accommodating 'minilateral' negotiations within a multilateral framework as an application of critical theory to practical negotiation. The proposal would be to replace the current *ad hoc* arrangements of the UNFCCC with a small Climate Council based on 'common but differentiated representation' including the most capable, the most responsible and the most vulnerable. It would be based on existing negotiating groups and could be as small as 8 members or as large as 23.

References

Abbott, K.W. (2012) 'The Transnational Regime Complex for Climate Change', *Environment and Planning C: Government and Policy*, 30(4): 571–90.

Adelle, C. and Withana, S. (2010) 'Public Perceptions of Climate Change and Energy Issues in the EU ,and the United States' in S. Oberthür and Pallemaerts, M. (eds) *The New Climate Policies of the European Union*, (Brussels: VUB Press): 309–35.

Adger, W.N. (2010) 'Climate Change, Human Well-Being and Insecurity', *New Political Economy*, 15(2): 275–92.

Adger, W.N., J. Paavola and S. Huq (2006) 'Towards Justice in Adaptation to Climate Change' in W.N. Adger, J. Paavola, S. Huq and M. Mace (eds) *Fairness in Adaptation to Climate Change* (Cambridge, MA: MIT Press): 1–19.

Aggarwal, V.K. (1998) *Institutional Design for a Complex World: Bargaining Linkages and Nesting* (Ithaca: Cornell University Press).

ALBA (2009) 'Declaration on Copenhagen Climate Summit', 28 December, venezuealanalysis.com/print/5083. Date accessed 12 May 2014.

Altman, R.C. and R.N. Haas (2010) 'American Profligacy and American Power: The Consequences of Fiscal Irresponsibility', *Foreign Affairs*, 89(6): 25–34.

Andonova, L.B., M.M. Betsill and H. Bulkely (2009) 'Transnational Climate Governance', *Global Environmental Politics*, 9(2): 52–73.

Asuka, J. (2014) 'Japan's Failing Climate Change Diplomacy', in Hoffmeister W. and P. Rueppel (eds) *EU Asia Dialogue: Climate Change Diplomacy: The Way Forward for Asia and Europe* (Singapore: Konrad Adenauer Stiftung and European Union): 23–36.

Atteridge, A., M.K. Shrivastava, N. Pahuja and H. Upadhyay (2012) 'Climate Policy in India: What Shapes International, National and State Policy?', *Ambio*, 41: 68–77.

Australian (2014) 'EU "Unhappy" Climate Change is off G20 Agenda', 3 April.

Australian Government Department of Climate Change and Energy Efficiency (2011) 'Feature: Cartagena Dialogue for Progressive Action' *Annual Report 2010-11*, www.climatechange.gov.au/about/accountability/annual-reports. Date accessed 23 May 2013.

Baechler, G. (1998) *Violence through Environmental Discrimination: Causes, Rwanda Arena and Conflict Model* (Dordrecht: Kluwyer).

Bali Principles of Climate Justice (2002) http://www.indiaresource.org/issues/energycc/2003/baliprinciples.html. Date accessed 23 May 2014.

Barnett, J. (2001) *The Meaning of Environmental Security: Ecological Politics and Policy in the New Security Era* (London: Zed Books).

BASIC (2013) 14th Ministerial Meeting, Beijing, 7 February.

BASIC (2013a) 16th Ministerial Meeting, Brazil, September 2013.

Basiur, R. (2011) 'India: A Major Power in the Making' in Volgy et al. *Major Powers* op.cit: 180–202.

Baylis, J., Smith, S. and Owens, P. (eds) (2013) *The Globalization of World Politics: An introduction to international relations*, sixth edition (Oxford: Oxford University Press).

Betzold, C. (2010) ' "Borrowing" Power to Influence International Negotiations: AOSIS in the Climate Change Regime, 1990–1997', *Politics*, 30(3): 131–48.

Biello, D. (2009) 'Can Climate Change Cause Conflict? Recent History Suggests So', *Scientific American*, 23 November: 1–3.

Biermann, F. et al. (2012) 'Transforming Governance and Institutions for Global Sustainability: Key Insights from the Earth System Governance Project', *Current Opinion in Environmental Sustainability*, 191: 1–10.

Biermann, F., P. Pattberg, H. van Asselt and F. Zelli (2009) 'The Fragmentation of Global Governance Architectures: A Framework for Analysis', *Global Environmental Politics*, 9(4) November: 14–40.

Bodansky, D. (1993) 'The United Nations Framework Convention on Climate Change: A Commentary', *Yale Journal of International Law*, 18: 451–558.

Bodansky, D.(2001) 'The History of the Global Climate Change Regime' in U. Luterbacher, and D. Sprinz (eds) *International Relations and Global Climate Change*, (Cambridge MA:MIT Press); 213–40.

Borger, J. (2009) 'David Miliband: China Ready to Join US as World Power', *Guardian*, 17 May.

Brennan, R. and Pettit, P. (2004) *The Economy of Esteem* (Oxford: Oxford University Press).

Brenton, T. (1994) *The Greening of Machiavelli: The Evolution of International Environmental Politics* (London: RIIA/Earthscan).

Brenton, A. (2013) ' "Great Powers" in Climate Politics', *Climate Policy*, 13(5): 541–6.

Breslin, S. (2013) 'China and Global Order. Signalling Threats or Friendship?', *International Affairs*, 89(3): 615–34.

Bretherton, C. and J. Vogler (2006) *The European Union as a Global Actor* (2nd edn.) (London: Routledge).

Bretherton, C. and J. Vogler (2009) 'Past its Peak? The European Union as a Global Actor 10 Years After', in F. Laursen (ed.) *The EU as a Foreign and Security Policy Actor* (Dordrecht: Republic of Letters): 23–44.

Brown, C. (1992) *International Relations Theory: New Normative Approaches* (London: Harvester Wheatsheaf).

Brown, O., A. Hammill and R. McLennan (2007) 'Climate as the "New" Security Threat: Implications for Africa', *International Affairs*, 83(6): 1141–54.

Bulkeley, H. and H. Scroeder (2012) 'Beyond State and Non-state Divides: Global Cities and the Governing of Climate Change', *European Journal of International Relations*, 18(4): 743–66.

Bull, H. (1977) *The Anarchical Society: A Study of Order in World Politics* (New York: Columbia University Press).

Buzan, B. (2004) *From International to World Society* (Cambridge: Cambridge University Press).

Buzan, B., C.A. Jones and R. Little (1993) *The Logic of Anarchy: Neo-Realism to Structural Realism* (New York: Columbia University Press).

Buzan, B., O. Waever and J. de Wilde (1998) *Security: A New Framework for Analysis* (Boulder, CO: Lynne Rienner).

Byravan, S and S.C. Rajan (2013) 'Loss and Damage Claims in Climate Justice' *The Hindu*, 2 December.

Caney, S. (2011) 'Global Distributive Justice and the State' in A. Banai, M. Ronzoni and C. Schemmel (eds) *Social Justice, Global Dynamics: Theoretical and Empirical Perspectives* (London: Routledge): 3–25.

Carin, B. and Mehlenbacher, L. (2010) 'Constituting Global Leadership. Which Countries Need to be Around the Summit Table for Climate Change' *Global Governance*, 16(1): 21–37.

Carr, E.H. (1939) *The Twenty Years Crisis, 1919-39: An introduction to the Study of International Relations* (London: Macmillan).

Cass, L. (2005) 'Norm Entrapment and Preference Change: The Evolution of the European Union Position on International Emissions Trading' *Global Environmental Politics*, 5 (2): 1–23.

Christoff, P. (2010) 'Cold Climate in Copenhagen: China and the United States at COP15', *Environmental Politics*, 19(4) July: 637–56.

Clapp, J. (1998) 'The Privatization of Global Environmental Governance: ISO 14000 and the Developing World', *Global Governance*, 4: 295–316.

Climateactiontracker.org (2013) Date accessed 20 October 2013.

Commission of the European Union (2011) *A Roadmap for Moving to a Competitive Low Carbon Economy in 2050*. COM (2011)112 Final.

Cooper, R. (2004) *The Breaking of Nations: Order and Chaos in the Twenty-first Century* (London: Atlantic Books).

Crutzen, P.J. (2002) 'Geology of Mankind', *Nature*, 415(6867): 23.

Dai, X. and Z. Diao (2011) 'Towards a New World Order for Climate Change: China and the European Union's Leadership Ambition', in R.K.W. Wurzel and J. Connelly (eds) *The European Union as a Leader in International Climate Change Politics* (London: Routledge): 252–68.

Depledge, J. (2006) 'The Opposite of Learning: Ossification in the Climate Change Regime', *Global Environmental Politics*, 6 (1) February: 1–22.

Depledge, J. (2008) 'Striving for No: Saudi Arabia in the Climate Change Regime', *Global Environmental Politics*, 8(4) November: 9–35.

Depledge, J. and F. Yamin (2009) 'The Global Climate-change Regime: A Defence', in D. Helm and C. Hepburn (eds) *The Economics and Politics of Climate Change* (Oxford: Oxford University Press): 433–53.

Dessler, D. (1989) 'What's at Stake in the Agent-structure Debate?' *International Organization*, 43(3) Summer: 441–73.

Detraz, N .and M.M. Betsill (2010) 'Climate Change and Environmental Security: For Whom the Discourse Shifts', *International Studies Perspectives*, 10(3): 303–20.

Deudney, D.H. (1990) 'The Case Against Linking Environmental Degradation and National Security', *Millennium*, 19(3): 461–76.

Deudney, D.H. and R.A. Matthews (eds) (1999) *Contested Grounds: Security and Conflict in the New Environmental Politics* (New York: SUNY Press).

Dimitrov, R.S. (2010) 'Inside UN Climate Change Negotiations: The Copenhagen Conference', *Review of Policy Research*, 27(6): 795–821.

Dobson, A. (2005) 'Globalization, Cosmopolitanism and the Environment', *International Relations*, 19(3): 259–73.

Dobson, A. (2006) 'Thick Cosmopolitanism', *Political Studies*, 54: 165–84.

Doherty, B. and T. Doyle (2013) *Environmentalism, Resistance and Solidarity* (Houndmills: Palgrave Macmillan).

Dorling, D. (2013) *Population 10 Billion* (London: Constable).

Drezner, D. (2007) 'The New World Order', *Foreign Affairs*, 86(2): 34–6.

Dryzek, J.S. (1997) *The Politics of the Earth: Environmental Discourses* (Oxford: Oxford University Press).

Eckersley, R. (2004) *The Green State: Rethinking Democracy and Sovereignty* (Cambridge, MA: MIT Press).
Eckersley, R. (2012) 'Moving Forward in the Climate Negotiations: Multilateralism or Minilateralism?', *Global Environmental Politics*, 12(2) May: 24–42.
Ehrlich, P.R. (1968) *The Population Bomb* (New York: Sierra Club/Ballantine Books).
Elliott, L. (2006) 'Cosmopolitan Environmental Harm Conventions', *Global Society*, 20(3): 346–63.
Emmott, S. (2013) *Ten Billion* (Harmondsworth: Penguin).
ENB (1995) 'Summary of the First Conference of the Parties for the Framework Convention on Climate Change: 28 March-7 April 1995' *Earth Negotiations Bulletin*, 12 (21): 1–11.
ENB (1997) 'Report of the Third Conference of the Parties to the United Nations Framework Convention on Climate Change. 1-11 December 1997' *Earth Negotiations Bulletin*, 12(76): 1–16.
ENB (2007) 'Summary of the Thirteenth Conference of the Parties to the United Nations Framework Convention on Climate Change and Third Meeting of the Parties to the Kyoto Protocol: 3-15 December 2007' *Earth Negotiations Bulletin*, 12 (354): 1–22.
ENB (2009) 'Summary of the Bonn Climate Change Talks: 1–12 June, Subsidiary Body 30', *Earth Negotiations Bulletin*, 12(241): 1–26.
ENB (2010) 'Summary of the Cancun Climate Change Conference: 29 November–11 December, COP 16', *Earth Negotiations Bulletin*, 12(148): 1–30.
ENB (2011) 'Summary of the Durban Climate Change Conference: 28 November–11 December, COP 17', *Earth Negotiations Bulletin*, 12(534): 1–31.
ENB (2012) 'Summary of the Doha Climate Change Conference: 26 November–8 December, COP 18', *Earth Negotiations Bulletin*, 12(567): 1–30.
ENB (2013) 'Summary of the Warsaw Climate Change Conference: 11–23 November, COP 19', *Earth Negotiations Bulletin*, 12(594): 1–32.
ENB (2014) 'Summary of the Lima Climate Change Conference: 1–14 December 2014, COP 20', *Earth Negotiations Bulletin*, 12(619): 1–46.
European Council (2008) *Climate Change and International Security: Paper from the High Representative and the Commission to the European Council*, S113/08.
European Council (2014) *Conclusions on the 2030 Climate and Energy Policy Framework, European Council (23 and 24 October 2014)*, SN 79/14.
European Council for Foreign Relations (2012) *Foreign Policy Scorecard*, www.ecfr.eu//scorecard.home. Date accessed 20 June 2013.
European Environment Agency (2012) *Climate Change Impacts and Vulnerability in Europe 2012; An Indicator Based Report*, no.12/2012 (Copenhagen: European Environment Agency).
European Parliament (2013) *Policy Briefing: EU and Russian Policies on Energy and Climate Change*, DG Expo/B/POLDEP/Note/2013_308.
European Parliament (2013a) *The Development of Climate Negotiations in View of Warsaw*, DG IP/A/ENVI/ST/2013-22, PE 507.493, October.
European Union (2013) *Submission by Lithuania and the European Commission on Behalf of the European Union and its Member States: The Scope, Design and Structure of the 2015 Agreement*. Vilnius, 16 September 2013, www.europa.eu. Date accessed 16 February 2014.

European Union (2013a) *Submission by Lithuania and the European Commission on Behalf of the European Union and its Member States: Further Elaboration of a Step Wise Process for Ambitious Mitigation Commitments in the 2015 Agreement.* Vilnius, 16 September 2013, www.europa.eu Date accessed 16 February 2014.

Evans, T. (1998) 'Doing Something Without Doing Anything: International Environmental Law', in T. Jewell and L. Steele (eds) *Law in Environmental Decision-Making* (Oxford: Clarendon Press): 207–27.

Falkner, R. (2005) 'American Hegemony and the Global Environment', *International Studies Review*, 7: 585–9.

Falkner, R. (2008) *Business Power and Conflict in International Environmental Politics* (Houndmills: Palgrave Macmillan).

Falkner, R. (2012) 'Global Environmentalism and the Greening of International Society', *International Affairs*, 88(3) May: 503–23.

Falkner, R., H. Stephan and J. Vogler (2010), 'International Climate Policy after Copenhagen: Towards a "Building Blocks" Approach', *Global Policy*, 1(3) October: 252–62.

Foot, R.(2006) 'Chinese Strategies in a US Hegemonic Global Order', *International Affairs*, 82(1): 77–94.

French Government (2015)'Paris 2015. For a Universal Climate Agreement', www.Cop21.gouv.fr/en/Cop21-cmp11/organisers. Accessed 15 June 2015.

Froggatt, A. and M.J. Levi (2009) 'Climate and Energy Security Policies and Measures: Synergies and Conflicts', *International Affairs*, 85(6): 1129–41.

Frost, M. (2006) *Global Ethics: Anarchy, Freedom and International Relations* (London: Routledge).

G8 (2009) *Responsible Leadership for a Sustainable Future*, Declaration, Aquila, Italy, 6 April, para 65.

G20 (2009) *London Summit- Leaders' Statement*, 2 April.

Garrett, G.(2010) 'G2 in 20. The United States and the World After the Global Financial Crisis', *Global Policy* 1(1): 29–39.

Galtung, J. (1964) 'A Structural Theory of Aggression', *Journal of Peace Research*, 1(2): 95–119.

Giddens, A. (1984) *The Constitution of Society: Outline of the Theory of Structuration* (Basingstoke: Macmillan).

Gillis, J. and Fountain, H. (2014) 'Trying to Reclaim Leadership on Climate Change' *New York Times*, 1 June.

Global Carbon Project (2014), *Global Carbon Atlas*, www.globalcarbonatlas.org/?=en/emissions.

Global Commons Institute (2011) *GCI Memo to UNFCCC Regarding Decisions at COP 17 for 'Increased Ambition'. Increased Ambition = Accelerated Convergence*, http://www.gci.org.uk. Date accessed 2 June 2014.

Goffmann, E.I. (1986) *Frame Analysis: An Essay on the Organization of Experience* (Boston: North East University Press).

Gomez-Echeverri, L. and B. Müller (2009) *The Financial Mechanism of the UNFCCC: A Brief History*, European Capacity Building Initiative Policy Brief, www.eurocapacity.org. Date accessed 2 February 2014.

Goron, C. (2014) 'EU-ASEAN Relations in the Post-2015 Climate Regime: Exploring Pathways for Top-down and Bottom-up Climate Governance' in Hofmeister, W. and P. Rueppel (eds) *EU-Asia Dialogue, Climate Change Diplomacy: The Way Forward for Asia and Europe* (Singapore: Konrad Adenauer Stiftung and European Union): 101–25.

Grieco, J. (1988) 'Anarchy and the Limits of Cooperation: A Realist Critique of the Newest Liberal Institutionalism' *International Organization*, 42 (August): 485–507.
Grubb, M. (1995) 'The Berlin Conference :Outcome and Implications' Briefing Paper no. 21 (London: RIIA).
Grubb, M. (2010) 'Copenhagen: Back to the Future', *Climate Policy*, 10(2): 127–30.
Guardian (2009) 16 November: 4.
Guardian (2013) 22 March: 1.
Haas, P.M. (1990) 'Obtaining International Environmental Protection Through Epistemic Communities', *Millennium*, 19(3): 347–364.
Hardin, G. (1968) 'The Tragedy of the Commons', *Science*, 162(3859): 1243–1248.
Hardin, G. (1974) 'Lifeboat Ethics: The Case Against helping the Poor', *Psychology Today*, September, www.garretthardinsociety.org/articles/art_lifeboat_ethics_case. Date accessed 14 June 2013.
Harris, P.G. (ed.) (2000) *Climate Change and American Foreign Policy* (New York: St Martin's Press).
Harris, P.G. (ed.) (2011) *Ethics and Global Environmental Policy: Cosmopolitan Conceptions of Justice* (Cheltenham: Edward Elgar).
Harris, P.G. (ed.) (2014) *Routledge Handbook of Global Environmental Politics* (Abingdon: Routledge).
Harvey, F. (2014) 'Failure to Reach a Deal in Paris would Fatally Undermine UN Efforts, Says European Chief', *Guardian*, 29 December: 11.
Hastings Dunn, D. and M.J.L. McClelland (2013) 'Shale Gas and the Revival of American Power: Debunking or Decline?', *International Affairs*, 89(6) November: 1411–28.
Hayward, T. (2007) 'Human Rights Versus Emissions Rights: Climate Justice and the Equitable Distribution of Ecological Space', *Ethics and International Affairs*, 21(4): 431–50.
Henkin, L. (1979) *How Nations Behave: Law and Foreign Policy* (2nd edn.) (New York: Columbia University Press).
Herman, P.F. and G.F. Treverton (2009) 'The Political Consequences of Climate Change', *Survival*, 51(2) April–May: 137–48.
Herz, M. (2011) 'Brazil: A Major Power in the Making?' in Volgy et al. *Major Powers* op.cit: 159–79.
Hochstetler, K. and E. Viola (2012) 'Brazil and the Politics of Climate Change: Beyond the Global Commons', *Environmental Politics*, 21(5): 753–71.
Homer-Dixon, T. (1999) *The Environment, Scarcity and Violence* (New Jersey: Princeton University Press).
Hone, D. (2013) '1965 Climate Change and Geoengineering' blogs.shell.com . Date accessed 16 February 2015.
Hopf, T. (1998) 'The Promise of Constructivism in International Relations Theory', *International Security*, 23(1): 171–200.
House of Commons Energy and Climate Committee (2012) *Twelfth Report: Consumption Based Emissions Reporting*, 18th April, www.parliament.uk/business/committees-a-z/commons-select/.../consumption-published. Date accessed 26 June 2013.
House of Commons Science and Technology Committee (2010) *Fifth Report of Session 2009–10* (London: UK Stationery Office).
Hovi, J and D.F. Sprinz (2006) 'The Limits of the Least Ambitious Program', *Global Environmental Politics*, 6(3) July: 28–42.

Hulme, M. (2010) 'Moving Beyond Climate Change', *Environment*, 53(3): 15–18.
Humphreys, D. (2011) 'Smoke and Mirrors: Some Reflections on the Science and Politics of Geoengineering', *Journal of Environment and Development*, 20(2): 11–20.
Hurrell, A. (2007) *On Global Order: Power, Values and the Constitution of International Society* (Oxford: Oxford University Press).
Hurrell, A and B. Kingsbury (eds) (1992) *The International Politics of the Environment* (Oxford: Clarendon Press).
Hurrell, A and S. Sengupta (2012) 'Emerging Powers and Global Climate Politics', *International Affairs*, 88(3) May: 463–84.
Ikenberry, G.J. (2001) *After Victory: Institutions, Strategic Restraint and the Rebuilding of Order After Major Wars* (New Jersey: Princeton University Press).
India Resource Center (2003) *Bali Principles of Climate Justice*, http://www.indiaresource.org./issues/energycc/2003/baliprinciples.html. Date accessed 23 February 2015.
International Chamber of Shipping (2012) *Shipping, World Trade and the Reduction of CO_2 Emissions*, London: International Chamber of Shipping, www.ics-shipping.org. Date accessed 14 May 2013.
International Energy Agency (2004) *World Energy Outlook* (Paris: IEA).
International Institute for Applied Systems Analysis (2011) *Climate Change Threats: Russia Faces Tough Climate Change Challenges*, 18 Options, Summer 2011, www.iasa.ac.at. Date accessed 14 May 2013.
IPCC (2013) 'Summary for Policymakers', in *Climate Change 2013. The Physical Science Basis. Contribution of Working Group I to the Fifth Assessment Report of the Intergovernmental Panel on Climate Change* (Cambridge: Cambridge University Press).
IPCC (2014) 'Summary for Policymakers', in *Climate Change 2014, Mitigation of Climate Change. Contribution of Working Group III to the Fifth Assessment Report of the Intergovernmental Panel on Climate Change* (Cambridge: Cambridge University Press).
IPCC (2014a) 'Chapter 13, International Cooperation Agreements and Instruments', in *Climate Change 2014. Mitigation of Climate Change. Contribution of Working Group III to the Fifth Assessment Report of the Intergovernmental Panel on Climate Change* (Cambridge: Cambridge University Press).
Jacques, P.J. (2012) 'A General Theory of Climate Denial', *Global Environmental Politics*, 12(2) May: 9–23.
Janicke, K.(2010) 'ALBA Summit in Venezuela Vows to Fight Climate Change with System Change' *Register*, http://londonprogressive journal.com/article/682. Date accessed 3 June 2011.
Jarju, P. (2014) 'Climate Change has Highest Political Importance for Poor Countries', *Climate Knowledge and Development Network*, 17 January: 3.
Jeffery, L, J. Gütschow, M. Viewig, M. Schaeffer, B. Hare, M. Rocha and F. Fallasch (2013) 'Japan: From Frontrunner to Laggard' *Climate Action Tracker: Policy Brief*, 15 November.
Jervis, R. (1991/92) 'The Future of World Politics: Will it Resemble its Past?', *International Security*, 16(3) Winter: 39–73.
Jinnah, S. (2011) 'Climate Change Bandwagoning: The Impacts of Strategic Linkages on Regime Design, Maintenance and Death', *Global Environmental Politics*, 11(3) August: 1–9.

Kahler, M. (2013) 'Rising Powers and Global Governance: Negotiating Change in a Resilient Status Quo', *International Affairs*, 89(3) May: 711–29.

Kahn Ribeiro, M. Kobayashi et al. (2007) 'Transport and its Infrastructure', in *Climate Change 2007. Mitigation. Contribution of Working Group III to the Fourth Assessment Report of the Intergovernmental Panel on Climate Change* (Cambridge: Cambridge University Press): 325–80.

Kanie, M. (2011) 'Japan as an Underachiever: Major Power Status in Climate Change Politics', in T.J. Volgy, R. Corbetta, K.A. Grant and R.G. Baird (eds) *Major Powers and the Quest for Status in International Politics: Global and Regional Perspectives* (New York: Palgrave Macmillan): 115–32.

Kanitkar, T., T. Jaayarman, M. De Souza, M. Sanwan, P. Purkayastha, R. Talwar and D. Raghunandan (2010) *Global Carbon Budgets and Burden Sharing in Mitigation –Complete Report* (Mumbai: Center for Science Technology and Society, Tata Institute for Social Sciences).

Karas, J. and T. Bosteels (2005) *OPEC and Climate Change: Challenge and Opportunities* (London: RIIA).

Karlsson, C., C. Parker, M. Hjerpe and B. Linnér (2011) 'Looking for Leaders: Perceptions of Climate Change Leadership Among Climate Change Negotiation Participants', *Global Environmental Politics*, 11(1) February: 89–107.

Keating, D. (2013) 'Aviation Emissions Deal in Doubt Despite EU Offer', *European Voice*, 19–25 September, 19(32): 1.

Keating, D. (2014a) 'Leaders Delay Decision on Climate Targets', *European Voice*, 27 March- 2 April, 20(12): 8.

Keating D. (2014b) 'Will MEPs Bow to Pressure on ETS?', *European Voice*, 13–19 March, 20(10): 11.

Keating, D. (2014c) 'Anatomy of a Climate Deal', *European Voice*, 6–12 November, 20(39): 13.

Keating, D. (2015) 'Cheap Oil Price Threatens EU Energy Plans', *European Voice*, 22–28 January, 21(3): 1–2.

Keohane, R.O. (1984) *After Hegemony: Cooperation and Discord in the World Political Economy* (New Jersey: Princeton University Press).

Keohane, R.O. (2010) 'The Economy of Esteem and Climate Change', *St Anthony's International Review*, 5(2): 16–28.

Keohane, R.O. and J.S. Nye (1977) *Power and Interdependence: World Politics in Transition* (Boston: Little Brown & Co).

Keohane, R.O. and D.G. Victor (2010) *The Regime Complex for Climate Change*, Discussion Paper 10-33, Belfer Center for Science and International Affairs, Harvard Kennedy School of Government, January.

Khan, M.H.I, C. Schwarte and S.T. Zaman (2013) 'Compensation for Loss and Damage', *Climate Justice Policy Brief*, Issue 1 (Dhaka: Centre for Climate Justice-Bangladesh).

King, D.A (2004) 'Climate Change Science: Adapt, Mitigate or Ignore?', *Science*, 303(9) January: 176–7.

Kirby, A. (2013) 'Least Developed Countries Agree to Cut Greenhouse Gas Emissions', *Guardian Environment Network*, Wednesday April 3.

Kokorin, A. and A. Korppoo (2013) *Russia's Post-Kyoto Climate Policy: Real Action or Window Dressing?*, ENI Climate Policy Perspectives 10 (Lysaker: Fridtjof Nansen Institute).

Krasner, S.D. (1983) ed. *International Regimes*, (Ithaca N.Y.: Cornell University Press).
Krasner, S.D. (1985) *Structural Conflict: The Third World Against Global Liberalism*,(Berkeley: University of California Press).
Kreft, S. and D. Eckstein (2013) *Global Climate Risk Index 2014*, Germanwatch, www.germanwatch.org/en/cri. Date accessed 14 May 2014.
Kutting, G. (ed.) (2011) *Global Environmental Politics: Concepts, Theories and Case Studies* (Abingdon: Routledge).
La Rovere, E.L., L.V. de Macedo and K.A. Baumert (2002) 'The Brazilian Proposal on Relative Responsibility for Global Warming', in K.A. Baumert, O. Blanchard, S. Llosa and J. Perkhaus (eds) *Building on the Kyoto Protocol: Options for Protecting the Climate* (Washington: World Resources Institute): 157–72.
Li, C and L. Xu (2014) 'Chinese Enthusiasm and American Cynicism: the "New Type of Great Power Relations"', *Brookings Institution: China-US Focus*, 4 December, Chinausfocus.com. Date accessed 14 May 2014.
Liebermann, B. and Schaefer, B.D.(2007) *Discussing Global Warming in the Security Council: Premature and a Distraction from More Pressing Crises* (Washington: Heritage Foundation).
Lietzmann, K.M. and G. Vest (eds) (1999) *Environment and Security in an International Context* (Brussels: NATO Committee on the Challenges of Modern Society).
Lindemann, T. (2010) *Causes of War: The Struggle for Recognition* (Colchester: European Consortium for Political Research Press).
Linklater, A. (2011) *The Problem of Harm in World Politics* (Cambridge: Cambridge) University Press.
Livemint (2014) 'Developing Countries Criticize IPCCC Report' 21 April, www.livemint.com/Politics. Date accessed 14 May 2014.
LMDC (2013) *Opening Statement for ADP COP 19*.
Lomborg, B. (2001) *The Skeptical Environmentalist: Measuring the Real State of the World* (Cambridge: Cambridge University Press).
Lynas, M. (2009) 'How do I know China Wrecked the Copenhagen Deal? I was in the Room', *Guardian*, 22 December.
Madzwamuse, M. (2014) *Drowning Voices: Climate Change Discourse in South Africa*, Heinrich Böll Stiftung Southern Africa, www.za.boell.org. Date accessed 16 May 2014.
Major Economies Forum (2013) *Chair's Summary, Sixteenth Meeting of Leaders' Representatives*, New York, 24 September.
Manners, I.(2002)'Normative Power Europe: A Contradiction in Terms?', *Journal of Common Market Studies,* 40(2): 235–58.
Marcoux, C. (2011) 'Understanding Institutional Change in International Environmental Regimes', *Global Environmental Politics,* 11(3) August: 145–51.
Mary Robinson Foundation (2011) *Principles of Climate Justice.*, www.mrfcj.org. Date accessed 10 June 2014.
McGrath M. (2014) 'IPCCC Scientists Accused of "Marginalising Poor Nations"' BBC News, 12 April.
Meltzer, J.(2011) 'After Fukushima: What's Next for Japan's Energy and Climate Change Policy?' *Global Economy and Development at Brookings* , (Washington: Brookings Institution).
Mitrany, D. (1975) *The Functional Theory of Politics* (London: LSE/Martin Robertson).

Mohan, C.J. (2010) 'Rising India: Partner in Shaping the Global Commons?, *Washington Quarterly*, 33(3): 133–48.
Morales, A. (2013) 'US, EU, Reject Brazilian Call for Climate Equity Metric', Bloomberg, http://www.bloomberg.com/news/print/3013-11-15. Date accessed 26 May 2014.
Morgenthau, H.J. (1967) *Politics Among Nations: The Struggle for Power and Peace* (New York: Alfred A. Knopf).
Motaal, D.A. (2010) 'The Shift from "Low Politics" to "High Politics": Climate Change' *Environmental Policy and Law*, 40(2–3): 98–109.
Nakićenović, N. et al. (2000) *Special Report on Emissions Scenarios: A Special Report of Working Group III of the Intergovernmental Panel on Climate Change* (Cambridge: Cambridge University Press).
Narlikar, A. (2006) 'Peculiar Chauvinism or Strategic Claculation? Explaining the negotiation Strategy of a Rising India', *International Affairs*, 82(1): 59–76.
Narlikar, A. (2013) 'Negotiating the Rise of New Powers', *International Affairs*, 89(3): 561–76.
Newell, P. (2012),*Globalization and the Environment: Capitalism, Ecology and Power* (Cambridge: Polity Press).
Nikiforuk, A. (2013) 'Oh Canada: How America's Friendly Northern Neighbour Became a Rogue Reckless Petro State, *Foreign Policy*, 24 June.
Nordhaus, T. and M. Shellenberger (2010) 'The End of Magical Climate Thinking', *Foreign Affairs*, 13 January: 19–28.
Oberthür, S. and M. Pallemaerts (2010) 'The EU's Internal and External Climate Policies: An Historical Overview', in S. Oberthür and M. Pallemaerts with C. Roche Kelly (eds) *The New Climate Policies of the European Union: Internal Legislation and Climate Diplomacy* (Brussels: VUB Press): 27–63.
Oberthür, S. and M. Pallemaerts with C. Roche Kelly (eds) (2010) *The New Climate Policies of the European Union: Internal Legislation and Climate Diplomacy* (Brussels: VUB Press).
Oberthür, S. and C. Roche Kelly (2007) 'EU Leadership in International Climate Policy: Achievements and Challenges', *The International Spectator*, 43(3): 35–50.
Okereke, C. and K. Dooley (2010) 'Principles of Justice in Proposals and Policy Approaches to Avoid Deforestation: Towards a Post-Kyoto Climate Agreement', *Global Environmental Change*, 20(1): 82–95.
O'Neil, B.C. (2009) 'Climate Change and Population Growth', in L. Mazur (ed.) *A Pivotal Moment: Population, Justice and the Environmental Challenge* (Washington, DC: Island Press): 81–94.
O'Neil, K. (2009) *The Environment and International Relations* (Cambridge: Cambridge University Press).
OPEC (2013) *Statement to the UN Climate Change Conference (COP 19)*, 22 November, www.opec.org. Date accessed 19 May 2014.
Opschoor, H. (2009) 'Sustainable Development and a Dwindling Carbon Space', *IIIS Public Lecture Series 2009* No.1 (The Hague: International Institute for Social Studies).
Organski, A.F.K. (1968) *World Politics* (2nd edn.) (New York: Knopf).
Ostrom, E. (1990) *Governing the Commons: The Evolution of Institutions for Collective Action* (Cambridge: Cambridge University Press).
Ott, H.E. (2002), 'Warning Signs from Delhi: Troubled Waters Ahead for Global Climate Policy', *Yearbook of International Environmental Law*, 13: 261–70.

Paterson, M.(1996) *Global Warming and Global Politics* (London: Routledge).
Pattberg, P. (2007) *Private Institutions and Global Governance: The New Politics of Environmental Sustainability* (Cheltenham: Edward Elgar).
Pearce F. (2011) *Peoplequake: Mass Migration, Ageing Nations and the Coming Population Crash* (London: Transworld).
Pearce, F. (2013) 'The Trillion-Ton Cap: Allocating the World's Carbon Emissions' *Yale Environment 360*, http://e360.yale.edu/content/print.msp?id=2703. Date accessed 3 February 2014.
Peters, G.P., J.C. Minx and C.L. Weber (2011) 'Growth in Emission Transfers via International Trade from 1990 to 2008', *Proceedings of the National Academy of Sciences*, 108(21): 8903–08.
Pickering, J., S. Vanderheiden and S. Miller (2013) '"If Equity's in, we're out": Scope for Fairness in the Next Global Climate Agreement', *Ethics and International Affairs*, 26: 423–43.
PricewaterhouseCoopers (2012) 'Reality Check for Rio Negotiations as PwC Analysis Examines Global Power Shifts to 2032', 19 June, http://pwc.blogs.compress_room/2012/06 Date accessed 10 July 2014.
Prins, G. et al. (2010) *The Hartwell Paper: A New Direction for Climate Policy after the Crash of 2009*, University of Oxford Institute for Science Innovation and Society/LSE Mackinder Programme, http://eprints.lse.uk/27939/ Date accessed 31 July 2014.
Purdon, M. (2013), 'Neo-classical Realism and International Climate Change Politics: Moral Imperative and Political Constraint in International Climate Finance', *Journal of International Relations and Development*, 16(1): 1–38.
Reid, T.R (2004) *The United States of Europe: The New Superpower and the End of American Supremacy* (London: Penguin).
Republic of South Africa (2011) *National Climate Policy Response White Paper* (Durban: Republic of South Africa).
Rifkind, J. (2004) *The European Dream: How Europe's Vision of the Future is Quietly Eclipsing the American Dream* (Cambridge: Polity Press).
Ringius, L. (1997) *Differentiation, Leaders and Fairness: Negotiating Climate Commitments in the European Community* (Oslo: Cicero, Report 1997): 8.
Roberts, T. and G. Edwards (2012) *A New Latin American Climate Negotiating Group: The Greenest Shoots in the Doha Desert*, Brookings, www.brookings.edu/blogs/up-front. Date accessed 16 May 2014.
Roberts, J.T. and B.C. Parks (2007) *A Climate of Injustice: Global Inequality, North-South Politics and Climate Policy* (Cambridge, MA: MIT Press).
Rowlands, I (1995)*The Politics of Global Atmospheric Change*(Manchester: Manchester University Press).
Royal Institute of International Affairs (1991) *Pledge and Review Processes: Possible Components of a Climate Convention: Workshop Report by Michael Grubb and Nicola Steen* (London: RIIA).
Royal Society (2009) *Geoengineering the Climate: Science, Governance and Uncertainty* (London: Royal Society).
Royal Society (2012) *People and the Planet*, The Royal Society Science Policy Report April 2012, DES2470 (London: Royal Society).
Saurin, J. (1996) 'International Relations, Social Ecology and the Globalisation of Environmental Change', in J. Vogler and M.F. Imber (eds) *The Environment and International Relations* (London: Routledge): 77–98.

Schapcott, R. (2010) *International Ethics: A Critical Introduction* (Cambridge: Polity).
Schroeder, H., M.T. Boykoff and L. Spiers,(2012) 'Equity and State Representation in Climate Negotiations', *Nature Climate Change*, 2, December: 834–6.
Schwarz, P. and D. Randall (2003) *An Abrupt Climate Change Scenario and its Implications for National Security* (San Francisco: Global Business Network).
Scrase, J.I. and D.G. Ockwell (2010) 'The Role of Discourse and Linguistic Framing in Sustaining High Carbon Energy Policy – An Accessible Introduction', *Energy Policy*, 38: 225–33.
Searle, J.R. (1995) *The Construction of Social Reality* (Harmondsworth: Penguin).
Sebenius, J.K. (1991) 'Designing Negotiations Towards a New Regime: The Case of Global Warming', *International Security*, 15(4) Spring: 110–48.
Seyfang, G, I. Lovenzoni and M. Nye (2009) *Personal Carbon Trading: A Critical Examination of Proposals for the UK*, Tyndall Centre for Climate Change Research, Working Paper 136.
Shue, H. (1992) 'The Unavoidability of Justice', in A. Hurrell and B. Kingsury (eds) *The International Politics of the Environment* (Oxford: Oxford University Press): 373–93.
Shue, H. (1993) 'Subsistence Emissions and Luxury Emissions', *Law and Policy*, 15(1): 39–59.
Shue, H. (1995) 'Ethics, the Environment and the Changing International Order', *International Affairs*, 71(3): 453–61.
Shue, H. (1999) 'Global Environment and International Inequality', *International Affairs*, 75(3): 531–45.
Skjaerseth, J.B.(1994) 'The Climate Policy of the EU: Too Hot to Handle', *Journal of Common Market Studies*, 32(1): 25–46.
Skolnikoff, E.B. (1990) 'The Policy Gridlock on Global Warming', *Foreign Policy*, 79, Summer: 77–93.
Skolnikoff, E.B. (1993) *The Elusive Transformation: Science, Technology and the Evolution of International Politics* (Princeton, NJ: Princeton University Press).
Smith, S. (1993) 'The environment on the periphery of international relations: an explanation', *Environmental Politics*, 2 (4): 28–45.
Smith, S. and P. Owens (2008) 'Alternative Approaches to International Theory', in Baylis, B., S. Smith and P. Owens (eds) *The Globalization of World Politics: An Introduction to International Relation* (4th edn.) (Oxford: Oxford University Press): 174–91.
South Centre (2011) *Operationalizing the UNFCCC Financial Mechanism*, Research Paper 39, Geneva: South Centre.
Spiegel Online (2008) 'WTO Failure Reflects Changing Global Power Relations' 30 July.
Sprinz, D. F and T. Vaahtoranta (1994) 'The Interest-based Explanation of International Environmental Policy', *International Organization*, 48(1): 77–105.
Stern, N. (2007) *The Economics of Climate Change: The Stern Review* (Cambridge: Cambridge University Press).
Stern, T.D. (2013) 'The Shape of a New International Climate Agreement', Transcript of a Speech Delivered at the Conference, 'Delivering Concrete Climate Change Action: Towards 2015', Chatham House, 22 October.
Stevenson, H. and J.S. Dryzek (2014) *Democratizing Global Climate Governance* (Cambridge: Cambridge University Press).

Stevis, D. (2006) 'The Trajectory of the Study of International Environmental Politics', in M.M. Betsill, K. Hochstetler and D. Stevis (eds) *Palgrave Advances in International Environmental Politics* (Basingstoke: Palgrave Macmillan): 13–53.

Strange, S. (1983) 'Cave! Hic Dragones: A Critique of Regime Analysis', in S.D. Krasner (ed.) *International Regimes* (Ithaca, NY: Cornell University Press): 337–54.

Strange, S. (1988) *States and Markets: An Introduction to International Political Economy* (London: Pinter Publishers).

Støre, J. (2010) Interview, *Spiegel on Line*, 22 June.

Talberg, A., S. Hui and K. Loynes (2013) *Australian Climate Change Policy: A Chronology* (Canberra: Parliament of Australia).

Tammen, R., J. Kugler and D. Lemke (2000) *Power Transition: Strategies for the 21st Century* (New York: Chatham House).

Tata Institute (2010) *Conference on Global Carbon Budgets and Equity in Climate Change:28–29 June 2010. Discussion paper, Supplementary Notes and Summary Report.*(Mumbai: Tata Institute of Social Sciences).

Terhalle, M. and J. Depledge (2013) 'Great-power Politics, Order, Transition and Climate Governance: Insights from International Relations Theory', *Climate Policy*, 13 (5): 572–88.

Thomas, D.S.G and C. Twyman (2004) 'Equity and Justice in Climate Change Adaptation Amongst Natural-resource-dependent Societies', *Global Environmental Change*, 15: 115–24.

Townsend, I. (2010) 'G20 at the November 2010 Seoul Summit', *House of Commons Library, SN/EP/5028.*

Underdahl, A. (1980) *The Politics of International Fisheries Management: The Case of the North-East Atlantic* (Oslo: Uniiversitetsforlaget).

UNEP (2011) *HFCs: A Critical Link in Protecting Climate and Ozone Layer,* unep.org/dewa/Portals/67/pff/HFC_report PDF. Date accessed 15 June 2013.

UNEP (2012) http://www.unep.org/ccac/About/tabid/101649/Default.aspx. Date accessed 15 June 2013.

UNEP (2013) *UN Environment Head Welcomes Signal to Combat Climate Change by the World's Two Largest Economies,* June 9, www.unep.org/newscentre. Date accessed 15 June 2013.

UNEP (2013a) *The Emissions Gap Report 2013: A UNEP Synthesis Report* (Nairobi: UNEP).

UNEP (2014) *UNEP Global Mercury Partnership,*www.unep.org/chemicalsandwaste/Mercury/Global Mercury Partnership.

UNFCCC (2009) *Least Developed Countries under the UNFCCC* (Bonn: Climate Change Secretariat).

UNFCCC (2011) *Guide for Presiding Officers,* unfccc.int/resource/docs/publications/guide_presiding_officers.pdf. Date accessed 25 January 2014.

UNFCCC (2013) *An Overview of the Mandates, as Well as the Progress of Work Under Institutions, Mechanisms and Arrangements Under the Convention, Note by the Secretariat,* ADP/2013/INF2, 30 October.

UNFCCC (2014) 'DRAFT TEXT on ADP 2-7 Agenda Item 3, Implementation of All Elements of Decision 1/CP.17' Version 1 of 8 December 2014.

United Nations (1975) *Yearbook of the United Nations 1972* (New York: United Nations Office of Public Information).

United Nations Department of Economic and Social Affairs (2010) *Environmental Indicators: GHGs*, http://esa.un.org/unstats.un.org Date accessed 20 June 2013.

United Nations Department of Economic and Social Affairs (2012), *World Population Prospects, the 2012 Revision*, http://esa.un.org/wpp/unpp/p2k0data.asp Date accessed 20 June 2013.

United Nations Security Council (2011) *6587th Meeting (AM & PM)*, SC/10332.

United States Council on Environmental Quality and Department of State (1981) *The Global 2000 Report to the President: Entering the Twenty First Century*, vol.1 (Charlottesville, VA: Blue Angel).

United States Executive Office of the President (2013) *The President's Climate Action Plan*, (Washington: The White House), www.whitehouse.gov Date accessed 12 February 2014.

United States Government (2014) 'US Submission on Elements of the 2015 Agreement', in E. King (ed.) 'US Outlines Plan for "Durable and Ambitious" UN Climate Deal' *Responding to Climate Change*, www.rtc.org Date accessed 1 December 2014.

United States White House (2014) *US-China Joint Announcement on Climate Chang, Beijing, China 12 November 2014*, www.whitehouse.gov Date accessed 1 December 2014.

University of Toronto (2011) G20 Information Centre.

US Embassy Cable, EO12958, 7 January 2010, 'Climate Change Provides GOC with Much Needed Distraction' http://www.guardian.co.uk/world/us-embassy cables-documents/ Date accessed 12 April 2012.

US Embassy Cable, EO 12958, 3 February 2010,'Ambassador Kennards Meeting with Spanish' http://www.guardian.co.uk/world/us-embassy cables-documents/ Date Accessed 12 April 2012.

US Embassy Cable, EO 12598, 17 February 2010. 'Deputy NSA Michael Froman visit to Brussels' http://www.guardian.co.uk/world/us-embassy cables-documents/ Date accessed 12 April 2012.

Van de Graaf, T. (2013) *The Politics and Institutions of Global Energy Governance* (Houndmills: Palgrave Macmillan).

van Schaik, L. (2012) *The EU and the Progressive Alliance Negotiating in Durban: Saving the Climate?*, ODI Working Paper 354 (London: Overseas Development Institute/Climate and Development Knowledge Network).

Victor, D.G. (2011) *Global Warming Gridlock: Creating More Effective Strategies for Protecting the Planet* (Cambridge: Cambridge University Press).

Victor, D.G., K. Raustiala and E.B. Skolnikoff (eds) (1998) *The Implementation and Effectiveness of International Environmental Commitments: Theory and Practice* (Cambridge MA:MIT Press).

Victor, D.G., M.G. Morgan, J. Apt, J. Steinbruner and K. Ricke (2009) 'The Geoengineering Option: A Last Resort Against Global Warming?', *Foreign Affairs*, 88: 64–76.

Vincent, R.J. (1986) *Human Rights and International Relations* (Cambridge: Cambridge University Press).

Vogler, J. (1999) 'The European Union as an Actor in International Environmental Politics', *Environmental Politics*, 8(3): 24–48.

Vogler, J. (2000) *The Global Commons: Environmental and Technological Governance* (2nd edn.) (Chichester: John Wiley).

Vogler, J. (2005) 'In Defense of International Environmental Cooperation', in J. Barry and R. Eckersley (eds) *The State and the Global Ecological Crisis* (Cambridge, MA: MIT Press): 229–54.

Vogler, J. (2011) 'EU Policy on Global Climate Change: The Negotiation of Burden-Sharing', in D.C. Thomas (ed.) *Making EU Foreign Policy: National Preferences, European Norms and Common Policies* (Houndmills: Palgrave): 150–74.

Vogler, J. (2012) 'Studying the Global Commons, Governance without Politics?', in P. Dauvergne (ed.) *A Handbook of Global Environmental Politics* (2nd edn.) (Cheltenham: Edward Elgar): 172–83.

Vogler, J. (2013) 'Changing Conceptions of Climate and Energy Security in Europe', *Environmental Politics*, 22(4) July: 627–45.

Vogler, J. and C. Bretherton (2006) 'The European Union as a Protagonist to the United States on Climate Change', *International Studies Perspectives*, 7: 1–22.

Vogler, J and Stephan, H.R. (2007) 'The European Union in Global Environmental Governance: Leadership in the Making? *International Environmental Agreements*, 7(4): 389–412.

Volgy, T.J. and A. Bailin (2003) *International Politics and State Strength* (Boulder: Lynne Reinner).

Volgy, T.J., R. Corbetta, K.A. Grant and R.G. Baird (eds) (2011) *Major Powers and the Quest for Status in International Politics: Global and Regional Perspectives* (New York: Palgrave Macmillan).

Waltz, K. (1979) *Theory of International Politics* (Reading, MA: Addison Wesley).

Ward, H. (1993) 'Game Theory and the Politics of the Global Commons', *Journal of Conflict Resolution*, 37(2): 203–35.

Ward, B. and D. Dubos (1972) *Only one Earth: The Care and Maintenance of a Small Planet* (Harmondsworth: Penguin).

Warner, K. and S.A. Zakieldeen (2011) *Loss and Damage Due to Climate Change: An Overview of the UNFCCC Negotiations*, European Capacity Building Initiative, www.eorocapacity.org. Date accessed 26 June 2013.

Webb, M.C. and S.D. Krasner (1989) 'Hegemonic Stability Theory: An Empirical Assessment', *Review of International Studies*, 15: 183–98.

Weber, M. (1948) 'The Social Psychology of World Religions', in H.H. Gerth and C.W. Mills (eds) *From Max Weber: Essays in Sociology* (London: Routledge and Kegan Paul): 267–301.

Wendt, A. (1994) 'Collective Identity Formation and the International State', *American Political Science Review*, 88(2): 384–96.

Wendt, A. (1999) *A Social Theory of International Politics* (Cambridge: Cambridge University Press).

Wight, M. (1978), *Power Politics* (2nd edn.), Hedley Bull and Carsten Holbrad (eds) (Harmondsworth: Penguin/RIIA).

Wolfers, A. (1962) *Discord and Collaboration* (Baltimore: Johns Hopkins University Press).

World Bank (2013) 'Data by Country', data.worldbank.org/country. World Coal Association (2015) 'Coal Statistics', www.worldcoal.org/resources/coal-statistics. Date accessed 6 February 2015.

World Commission on Environment and Development (1987) *Our Common Future* (Oxford: Oxford University Press).

World Trade Organization (2013) *Activities of the WTO and the Challenge of Climate Change*, http://www.wto.org/english/tratop_e/envir_e/climate_challenge_e.htm Date accessed 12 February 2014.

Wurzel, R. and J. Connelly, (eds) (2010) *The European Union as Leader in International Climate Change Policy* (London: Routledge).

Yamin, F. and J. Depledge (2004) *The International Climate Regime: A Guide to Rules, Institutions and Procedures* (Cambridge: Cambridge University Press).

Young. O.R. (2010)*Institutional Dynamics: Emergent Patterns in International Environmental Governance* (Cambridge MA: MIT Press).

Zakaria, T. (2008) *The Post-American World* (London: Allen Lane).

Zhang, Z. (2003) 'The Forces Behind China's Climate Change Policy: Interests, Sovereignty and Prestige', in P.G. Harris (ed.) *Global Warming and East Asia* (London: Routledge): 66–85.

Index

A
Abbot, Tony, 70, 72
acid rain, 61
actions, judgment of, 88
adaptation
  2020 agreement, 51
  importance of, 51
  meaning of, 50
  problems, 50
  provisions for loss and damage, 51, 59
  see also climate change
Ad hoc Group on the Berlin Mandate (AGBM), 38, 55, 134
Ad hoc Working Group of the Kyoto Protocol (AWG-KP), 55, 180n4
Ad hoc Working Group on Long-term Cooperative Action under the Convention (AWG-LCA), 38, 55, 180n4
Ad hoc Working Group on the Durban Platform for Enhanced Action (ADP), 38, 43, 48, 55, 181n12
African National Congress (ANC), 80
air freight, 26
air pollutants, 39
alignments in climate diplomacy, 62, 64
Alliance of Small Island States (AOSIS), 10, 40, 60, 64, 75, 82–3, 104, 106, 112, 139, 156
'All Quiet on the Western Front', anti-war film, 113
Anthropocene age, 24, 28, 179n10
anthropogenic emissions, 9
Asia Pacific Economic Cooperation (APEC), 18, 68–9, 125
Asia-Pacific Partnership on Clean Development and Climate, 32, 125
assigned amount units (AAUs), 46, 181n13

Association of Independent Latin American and Caribbean States (AILAC), 81–2, 84, 106, 168
Association of South East Asian Nations (ASEAN), 84

B
balance of power, 151
Bali Declaration of Principles of Climate Justice (2002), 86
Bali Plan of Action (2007), 41, 43, 51–2, 59, 91, 103, 129
Beckett, Margaret, 22
Berlin Mandate (1995), 41, 62, 139, 153
Bilateral Offset Crediting Mechanism (BOCM), 70
Bolivarian Alliance of the Peoples of Our America (ALBA), 11, 57, 81, 116–17, 124
Brazil, Russia, India, and China (BRICs), 135, 141–4, 184n2
Brazil, South Africa, India and China (BASIC) group, 10, 12, 46, 64, 77–81, 84, 93, 110, 117, 119, 124–8, 141–4, 150–1, 155
Bretton Woods monetary arrangements, 37
Brundtland Commission Report, 15
Bull, Hedley, 89
Bush, George W., 63, 68, 93
business as usual (BAU), 185n3
Byrd Hagel Resolution (1997), 68, 93–4, 182n4

C
Cancun Adaptation Framework (2010), 51
Cancun COP (2010), 40, 43, 47, 120
capitalism, 116–17, 143
Carbon Capture and Storage (CCS), 149

carbon dioxide, 29, 104
  China as major emitter
    since 2004, 41
  emissions, 79, 146–7
carbon dioxide removal (CDR), 28, 30
carbon emissions, 77, 101, 182n8
carbon-intensive industry offshore,
    displacement of, 27
carbon leakage, 27, 66
carbon space, 10, 99–100, 102, 104,
    106, 154
carbon tax, 94, 181n3
  industrial lobbying over, 62
carbon-trading, 27
Carr, Edward Hallett, 172, 173
Cartagena Dialogue, 10, 84, 106, 135,
    141, 156, 171
Castro, Fidel, 116–17
Certified Emission Reduction Units
    (CERs), 45, 51
Chavez, Hugo, 116
chlorofluorocarbons (CFCs), 19
Clean Development Mechanism
    (CDM), 45, 70, 97, 118
climate action, 52, 65, 80, 113, 140
climate activism, 70, 120, 169
climate agreement disillusionment
    with interstate cooperation, 2
Climate and Clean Air Coalition to
    Reduce Short-lived Pollutants
    (2012), 32
Climate and Energy Policy Framework
    (2030), 67
climate change, 1, 6, 10, 17, 33, 36,
    51, 86–8, 182n2
  distribution of economic and
    technological capabilities, 133
  international attempts to solve
    problem, 13
  measures to combat, 27
  negotiations, 10, 62, 69, 75, 77, 81,
    83, 90, 112, 130, 144, 151, 153,
    169–70, 173, 177, 181n2
  policy significance, 89
  politics of, 166–71
  and politics of prestige, 110–12
  and population, 24–6
  security of, 21–4
  social and economic changes
    impact on, 61–2

  see also Intergovernmental Panel on
    Climate Change (IPCC); Kyoto
    Protocol (1997); United Nations
    Framework Convention on
    Climate Change (UNFCCC)
climate convention, 16–21
Climate Convention (1992), 4, 27
climate diplomacy, 11
climate gridlock, 58
climate-induced loss and damage, 86
climate justice
  definition of, 86
  egalitarian form of, 91
climate policy leadership, 117–21
climate politics, 5, 8, 10–12, 33–4, 62,
    66, 86
climate regime, 2, 4, 7, 12, 15, 58, 89
  actual implementation of, 105–6
  changing context of, 133–44
  effectiveness of, 158–65
  future of, 166–71
  internal political system of, 9
  irrational behaviour within, 11
  JUSSCANZ coalition, 67
  macro rationale for, 105
  structural change, 131
  and trends, shifting relationship
    between, 8
  see also United Nations Framework
    Convention on Climate
    Change (UNFCCC)
climate rights and obligations, 90–1
climate vulnerability, 7, 10, 25, 28
  assessment of, 50
  neat indices of, 61
Cold War, 14, 21, 33, 123–4, 133,
    136–7, 145, 150
common but differentiated
  responsibilities and respective
  capabilities (CBDR-RC), 4, 10,
    19, 24, 32, 40–1, 44, 48, 87,
    92–4, 106, 137, 150, 180n7
common heritage, 17, 123
commons governance, 36–7
commons regimes, 36, 38
commons tragedies, 36
communitarian approach, 10–11
  position of pluralism, 89
communitarianism, 8, 88–9, 182n2
communitarian pluralist approach, 90

community status attribution, 112
compatriot priority, 88
conflicts over energy resources, 112
contraction and convergence model, 99, 102
Convention on the Prohibition of Military or any other Hostile Use of Environmental Modification Techniques (ENMOD) Treaty, 1977, 29
COP 7 Marrakesh Accords, 44
Copenhagen Accord, 38, 40, 52, 71, 109, 117, 119, 126, 129–30, 156, 181n2, 183n2, 184n10
  Green Climate Fund, 47
Copenhagen Conference of the Parties (COP) of 2009, 11, 43–4, 46, 54, 58, 77, 79, 81, 110, 119, 129, 131, 135, 143–4, 153, 168, 173
  ALBA actions/interactions at, 116–17, 169
  EU failure at, 141
corporate interests, 63
cosmopolitan/cosmopolitanism approach, 8, 10, 90, 101, 183n3

D
Darfur conflict, Sudan, 22
Delegation General (DG) for Climate Action, 65–6
democratic accountability, 87
Dimitrov, Radoslav, 117
distributive climate justice, 10
Dobson, Andrew, 90
Doha COP (2012), 43, 55, 71, 88, 92, 109, 180n4, 185n3
domestic aviation emissions, 21
Dryzek, John, 28
Durban Platform (2011), 38, 43, 59, 104, 109, 128–9, 141, 144, 156, 180n4, 185n3

E
Earth Summit (1992), 110, 124
Earth Systems Governance Project, 3
ecological debt, 87
ecological sustainability, 87
ecological vulnerability, 61
economic globalization, 133
economic interests, 60–1
economic sovereignty over natural resources, 115
economies in transition (EIT), 137
emerging powers, 123, 150
Emission Reduction Units (ERUs), 45
emissions, 137–8, 148
  exclusion, 8–9
  gap by 2020, 47
  by high income societies, 86
  historic cumulative (1850–2030), 97
  increase among developed countries, 41–2
  intensity targets, 93
  pattern influence on states, 133
  per capita distribution by country, 101
  reduction pledges, 71
  reduction targets, 93
  trading, 45, 59, 66, 94
  see also Kyoto Protocol (1997); United Nations Framework Convention on Climate Change (UNFCCC) 1992
Emissions Trading Scheme (ETS), 18, 66, 71, 118, 141
Energy Union, 149
environmental agreements, 49, 60, 159
environmental degradation, 27, 116, 154
Environmental Integrity Group (EIG), 10, 73–4
environmental politics, 2, 6, 44, 115, 132, 153, 158
Environmental Protection Agency (EPA), 69
environmental security, 23
environmental vulnerability, 61
equity, 87, 91–2, 96, 100–1
  in climate politics, India as champions, 124
ethical theory in international relations, 88–91
European Commission, 63, 117, 149
European Union (EU), 5, 8, 10, 41, 62, 138, 153, 155–6

achievements during Kyoto ratification process, 67
agreed to include international flights emissions, 21
associated with group of associated states, 65
burden-sharing agreement (2008), 94–5
classification in terms of environmental policy, 65–6
in climate change convention, 65
climate leadership, 114
role of, 11
in climate negotiations, 64
Copenhagen COP, 141
emissions 2012, 96
emissions reduction by 2020, 130
ending of climate diplomacy after Cold War, 139
energy supply structure, 149
enlargements of 2004 and 2007, 140
foreign relations, 65
high-level discussion of climate change regime, 125
indicators of structural change, 147
international climate policy leadership, 118
Kyoto Protocol negotiation, 66
leader in climate politics, 66
member states, 65
plans to decarbonise and set GHG reduction, 18
position on 2015 agreement, 67
pressed for comprehensive new protocol, 48
pronounced mean temperature rise in 1996, 40
reputation in climate politics, 128
steel-making capacity, 148
step-wise approach, 95, 180n9
Eurozone crisis, 141
EU–US climate relations, 46

F
fairness, concept of, 92–3
fair treatment, notion of, 93
Falklands/Malvinas conflict (1982), 108
Former Yugoslav Republic of Macedonia (FYROM), 65
fossil fuels, 18, 26, 45, 61–3
frame(s), analysis, 13
frameworks, types of, 14
fragmentation, 15
framing(s), concept of, 13
climate convention and Kyoto Protocol, 16–21
of climate problem, 14
internal structure of government, 15
part of ideological superstructure, 33
'Friends of the Chair Group' of 25, 110
Fukushima disaster, 71

G
G2 (United States and China), 131, 144
G7, 120, 122–6
G8+5, 124, 126
G8 Gleneagles summit (2005), 140
GDP growth for major economies, 145–6
General Agreement on Tariffs and Trade/World Trade Organisation (GATT/WTO) system, 27, 93, 155
Geneva Ministerial Declaration (1996), 57
geoengineering, 28–31
global civil society, 3
Global Climate Coalition, 62
Global Climate Fund, 53
Global Commons Institute (GCI), London, 99
global cooling, 16
global economic crisis/financial collapse (2007–8), 140–2
global environmental change, international relations of, 2
global environmental governance, 6
global environmental problem, state responsible for, 2
Global Environment Facility (GEF), 50–1, 180n10
criticism over operations, 52
global governance, 1, 14–16
global/international environmental politics, 6
private governance importance to study, 2
globalisation process, 26–8
Global South, 19

G

global warming, 2, 16, 19, 28, 90, 163
Goffmann, Erving, 13
Goldman Sachs, 142
Green Climate Fund (GCF), 47, 52, 182n7
greenhouse gases (GHGs) emissions mitigation, 4, 8, 10–11, 17–18, 30, 39, 44, 61, 67–8, 131, 140, 148
  and demographic change, relationship between, 25
  emitted by modern societies, 90
  international transport contribution to, 20
  measures to restrict, 27
  mitigation under climate convention, 19
  Obama announcement on net cut in, 69
  post-2020, 47–9
  review process importance, 50
  principles and rules, 44–7
  nationally defined contributions, 92–4
  per capita entitlement to global carbon budget, 98–102
  relative responsibility, 96–8
  targets and timetables, 94–6
  problem of, 178n5
  sectors contribution to, 20
  shipping sector responsible in, 21
  significance of historical concentrations of, 98
  South Korea doubling of, 74
  temporal dimension with, 87
green state, 3
Group of 8 (G8), 1, 18, 22, 31, 112, 122, 124, 185n3
Group of 20 (G20), 18, 31, 72, 112, 125–8, 142–4, 149, 185n3
Group of 77 (G77), 9–10, 23, 31, 32, 43, 46–7, 68, 74–7, 82–4, 93, 95, 112, 123–4, 129, 136–7, 139, 155–6, 180n2

H

Hardin, Garrett, 36
hegemonic leadership, 133, 152–4, 170
hegemonic stability, notion of, 133, 151
Helsinki Protocol (1985), 61
honour, 108
human consumption, 26–8
human immunodeficiency virus/ acquired immune deficiency syndrome (HIV/AIDS), 23
human security, 23
Hurricane Katrina, 68
hydrocarbons burning, climate problem due to, 17–18
hydrochlorofluorocarbons (HCFCs), 19, 33
hyrdofluorocarbons (HFCs), 19, 31–2, 78, 127, 151, 182n9

I

institutions
  created for functionalist purposes, 6
  functional for states, 6–7
intellectual property rights (IPR), 116
Intended Nationally Determined Contributions (INDCs), 48
interest-based bargaining, 61
interest, notion of, 60
Intergovernmental Negotiating Committee (INC), 37, 40, 48, 68, 88, 153, 155
Intergovernmental Panel on Climate Change (IPCC), 4–5, 5, 25–8, 32, 39–40, 54, 62, 99, 102, 118, 150, 178n3, 184n3, 185n4
  see also United Nations Framework Convention on Climate Change (UNFCCC) 1992
International Assessment and Review (IAR) process, 49
international aviation, 9
International Civil Aviation Organisation (ICAO), 20–1, 31, 141, 179n7

International Consultation and Assessment (ICA) process, 50
International Convention for the Prevention of Pollution from Ships (MARPOL), 21, 178n5
international cooperation, 3–4, 15, 26, 31
  liberals views on, 61
  problem for, 36
  UNFCC regime, 37
International Energy Agency (IEA), 18, 41
international environmental cooperation, 1, 11
  characteristics of, 6
  functions performed by nation-states, 3–4
  imposition of regulatory control by state authorities, 4
  role of, 4–5
international environmental negotiations, 11
international ethics
  distinction in treatment of, 88
  pluralist approach to, 89
international institutional architecture, 15
International Maritime Organisation (IMO), 20–1, 31
International Monetary Fund (IMF), 150, 154, 184n2
international political structure, 11
international politics, 85, 112, 128, 170, 173
  civilian dimensions, 120
  of climate change, 172
  importance of, 1
  prestige-seeking and demands for recognition, 168
international relations (IR), 8, 15, 37, 108, 178n2, 182n1
  agent structure debate, 132
  of climate change, 171–7
  ethical theory in, *see* Ethical theory in international relations
  functional approach to, 6
  study of identity and recognition, 11
interstate climate regime, 2
issue structure in climate regime, 154–5

J
Japan, US, Switzerland, Canada, Australia and New Zealand (JUSSCANZ) coalition, 67, 83, 139, 184n1
Jervis, Robert, 137
Jinbao, Wen, 127
Jinping, Xi, 78, 127, 178n4
justice, pursuit of, 86
  utilitarian conceptions of, 93

K
Kant, Immanuel, 90
Kellogg-Briand Pact, 106
Keohane, Robert O., 129
Krasner, Steven, 115
Kyoto Protocol (1997), 2, 8, 13, 16–21, 32, 41, 45, 53–4, 58–9, 63, 69–70, 75, 83, 87, 113, 144, 152–3, 157, 179n1, 184n2
  adaptation fund, 51
  Bush administration campaign against, 93
  emissions trading, 66
  equitable treatment concept, 94
  establishment of, 138–41
  Japan declines to join second phase EU, 121
  mitigation activities of developed countries, 9
  required mitigation action by developed countries, 106
  Russia defended hot air allocation, 61
  sources of credits, 46
  as a top down agreement, 44, 48
  US government withdrawal in 2001, 118
  see also United Nations Framework Convention on Climate Change (UNFCCC)

L

land-use, land-use change and forestry (LULUCF), 19, 46, 67, 185*n*3
Le Bow, Richard Ned, 108
League of Nations, 106
least developed countries (LDCs), 4, 10, 51, 63, 75–6, 82, 103, 179*n*1
lifeboat ethics, 25
like-minded developing countries, 10
like-minded group of developing countries (LMDC), 84–5
Linklater, Andrew, 90
Lisbon Treaty 2009, 181
Lomborg, Bjørn, 93
Long-range Transboundary Air Pollution (LRTAP) 1979, 39

M

Machiavelli, Niccolò, 88
Major Economies Forum (MEF), 122, 125, 185*n*3
Major Economies Meeting on Energy Security and Climate Change (2007), 125
Malthus, Thomas, 24
maritime emissions, 118, 179*n*5
marketisation, 28
Marrakesh Accord/Agreement, 27, 119
Marrakesh COP, 75
measurement reporting and verification (MRV), 49–50
Merkel, Angela, 55
Meyer, Aubrey, 89
military power, 146
Millennium Development Goals, 26
Mitrany, David, 6
monitoring, reporting and verification (MRV), 115–16, 183*n*2
Montreal Protocol (1985-7), 5, 17, 19, 30–1, 61, 151, 182*n*9
Moon, Ban Ki, 111
Morales, Evo, 117
Morgenthau, Hans J., 108, 128, 172

N

Nagoya Protocol (2010), 121, 183*n*6
National Adaptation Plans (NAPs), 51
national delegates to climate conferences, 63
national ecological vulnerability, 61
national interest
  actors as sovereign nation-states, 62
  in international relations, 64
  national delegates to climate conferences, 63
  nature of, 60
  US, 68
Nationally Appropriate Mitigation Actions (NAMAs), 41–5, 47, 50, 115, 183*n*2
  see also Copenhagen Accord
nationally defined contributions approach, 93
national recognition, 109
natural disasters, 157
natural frameworks, 14
neorealism, 132
New International Economic Order (NIEO), 74, 105, 115, 123, 136
New York climate summit (2014), 111
non-aligned movement, 123
non-governmental organisations (NGOs), 3, 39, 54, 62–3, 113–14, 150, 181*n*14
non-material interests, 109
Non-Proliferation Treaty, 124
North Atlantic Treaty Organisation (NATO), 22
North West Passage, 22

O

Obama, Barack, 69
ocean fertilisation, 29
'one child' policy, China, 26
O'Neill, Jim, 142
open market capitalism, 33
Organisation for Economic Cooperation and Development (OECD), 40, 62, 70, 73, 89, 136, 153
Organisation of Petroleum Exporting Countries (OPEC), 10, 76, 139, 149, 180*n*2
Ostrom, Elinor, 36
ozone layer, 5, 17, 19, 30

P

Paris agreement (2015), 43
Parties to the Convention, 9, 40
physical climate change, 13
Ping, Deng Xiao, 136
pluralism, 89, 106, 182n2, 183n4
policy frames and international
    institutional architecture, link
    between, 15
political action, 90
political structure, 132
possession and milieu goals,
    distinction between, 60
possession goals, 9–10
power politics, 108
power transition, 133
prestige, politics of, 128
  climate change and, *see* Climate
    change and politics of prestige
  community ascription,
    121–2
  contest for, 108
  implications of, 112
private governance, 2

Q

Quantified Emissions Limitation or
    Reduction Objectives (QELROs),
    44, 47, 94, 96

R

re-afforestation technique, 28
reality, scientific versions of, 14
recognition, 8, 11, 77, 109, 112
  and assertion of sovereignty,
    115–17
  North–South confrontation, 5
  politics of, 114
  significance of, 113
reducing emissions from deforestation
    and degradation in developing
    countries, including
    conservation (REDD+), 9, 53,
    80, 103, 181n11
Regional Economic Integration
    Organization (REIO), 62, 65
relational power, 132
rights, 86

Rio Earth Summit (1992), 15, 35, 38,
    41, 44, 117
Royal Institute of International Affairs
    (RIIA), 49

S

Saño, Nadrev, 157
scientific framing of climate
    problem, 32
sea transport, 9
Searle, John, 13
Seattle WTO ministerial (1999), 142
securitisation of climate, 21–4
Security Council and the General
    Assembly, Economic and Social
    Council (ECOSOC), 23–4, 32
self-esteem, 108
shale gas, 152
shipping containerisation, 26
Shue, Henry, 98
single national interest, notion of, 62
Small Island Developing States
    (SIDS), 51
social fact, 14
social frameworks, 14
social justice, 87
socioeconomic structure, 132
solar radiation management (SRM),
    28, 30
standing, 108
state boundaries, 88
status, politics of, 130
  attribution, 128
  emergent powers and quest for
    status, 122–5
  implications of, 112
Stern Review (2006), 36, 45, 61
Stern, Tod, 92
Stockholm Declaration (1972), 89,
    115, 155
Strange, Susan, 132, 148, 158
stratospheric ozone depletion, 14,
    17, 61
structure/structural, concept of
  framework of building or transport
    system, 132
  interpretations, 11
  power, 132, 151–4

structure/structural – *continued*
significance, 132
trends, 144–51
Subsidiary Body for Scientific and Technological Advice (SBSTA), 32–3, 39, 55–6, 96
sustainable development, 35, 42
bargain, 105
right of parties to, 88
symbolic politics, 11

T
Technology Executive Committee (TEC), 52
territorial emissions, 8
Thatcher, Margaret, 109, 134, 136, 138
tight oil, 152
trade liberalisation, 27
tragedies, 36
transborder pollution, 1
trans-boundary environmental governance, 36
transport emissions, global, 20
Treaty of Maastricht (1992), 138
Treaty on the Functioning of the European Union (TFEU), 181n2

U
Ukrainian gas crises (2006 and 2009), 149
UK Royal Society Report, 30
Umbrella Group, The, 10, 43, 48, 64, 67–73, 83, 92, 106, 124
see also Kyoto Protocol (1997)
UN Conference on Population and Development (1994), 26
UN Conference on the Human Environment (1972), Stockholm, 23–4
UN Convention on the Law of the Sea (UNCLOS) 1982, 15, 17
UN General Assembly (UNGA)
climate as a common concern, 17
climate convention, 19
United Nations Climate Change Convention, 2
United Nations Conference on the Human Environment (UNCHE) 1972, 16

United Nations Conference on Trade and Development (UNCTAD), 74
United Nations Development Programme (UNDP), 23
United Nations Environment Programme (UNEP), 19, 31–2, 129, 183n6
United Nations Framework Convention on Climate Change (UNFCCC) 1992, 1, 5, 7–8, 12–13, 15–16, 26, 31, 35, 70, 84–5, 92, 99, 118, 130, 137, 148–9, 151, 185n6
annual carbon-intensive gathering of useless frequent flyers process, 157
GHG mitigation under, 19, 26
history of, 9
initial framing of climate problem in, 17
institutional stasis of, 31
issue-related power structure, 155
mutual aid, parties duty of, 103–5
old Soviet bloc economies of, 72
parties responsibility for loss and damage, 103–5
principle of, 39
regimes, 59, 158
adaptation, *see* Adaptation
Berlin Mandate, 41
chronology of, 38
decision-making procedures, 54–8
dialogue with scientific findings, 39
equity principle, 40–1
finance aid provision, 51–3
measurement reporting and verification (MRV), *see* Measurement reporting and verification (MRV)
Nationally Appropriate Mitigation Actions (NAMAs), *see* Nationally Appropriate Mitigation Actions (NAMAs)
origin of, 37
provision for legal framework, 36
Rio Earth summit, 44
top down targets, 37

rights and duties, 87
US reengagement in Bali with, 69
UN Security Council, 1, 22, 32, 124
US President's Science Council report (1965), 29

V
Victor, David, 18
Vienna Convention (1985), 17, 39
Vincent, John, 89, 91
Volgy, Tom, 121

W
Walters, Sir Alan, 109
Waltz, Kenneth, 132

Warsaw COP (2013), 157
Warsaw Framework for REDD+, 53–4
Weber, Max, 34
Wight, Martin, 108, 172
Wolfers, Arnold, 60, 172
World Trade Organisation (WTO), 123–4, 131, 142
   approach to climate change, 28
   Committee on Trade and Environment, 28
   predictability of commitments, 28

Y
Young, Oran, 6

Made in the USA
Middletown, DE
22 January 2019